An Introduction to Real Analysis

Second Edition

Alan Parks
Lawrence University,
Appleton, Wisconsin

♠ This project puts text material, carefully coordinated with lectures, homework, and bibliography, into the hands of upper level students at cost. It was begun during the summer of 2003 and revised continuously since then.

©Alan Parks. All rights reserved. No portion of this work may be reproduced, stored in a retrieval system, or transmitted in any form, or by any means, electronic, mechanical, photocopying, recording, or otherwise, without the prior consent of the author. The principal copy of this work was printed on January 15, 2014.

Contents

Introduction.   v

Chapter 1.   The Real Numbers.   1
1. Completeness.   2
2. Rational and Irrational Density.   6
3. Problems   9

Chapter 2.   Sequences.   13
1. Definitions and Examples.   13
2. Convergence.   15
3. Monotone Convergence.   23
4. Problems   28

Chapter 3.   Limit Points and Function Limits.   31
1. Limit Points and BW.   31
2. Consequences of BW.   34
3. Function Limits.   37
4. Infinity.   44
5. Problems   46

Chapter 4.   Continuous Functions.   49
1. The Definition.   49
2. On a Closed Interval.   52
3. Inverses.   56
4. Uniform Continuity.   58
5. Problems   61

Chapter 5.   The Derivative.   65
1. Secant Functions.   65
2. Interior Extremes, Mean Values.   71

3. Inverse Functions. 74
  4. L'Hôpital's Rule. 76
  5. Problems 78

Chapter 6. The Definite Integral. 83
  1. Introduction to Integration. 83
  2. Variation Sums and Integrable Functions. 85
  3. Algebraic Combinations of Integrable Functions. 90
  4. Upper and Lower Sums. 95
  5. The Integral. 98
  6. Riemann Sums and Integral Algebra. 99
  7. The Fundamental Theorem of Calculus. 103
  8. Notation and Applications. 107
  9. Problems 114

Chapter 7. Uniform Convergence. 119
  1. The Definition 119
  2. Problems 123

Chapter 8. Taylor Series. 125
  1. Functions As Series. 125
  2. Radius of Convergence. 129
  3. Convergence, Continuity, Differentiation. 133
  4. Abel's Theorem on the Endpoints. 138
  5. Favorite Examples. 140
  6. Problems 144

Appendix A. The Real Numbers. 149
  1. Properties Except for Completeness. 149
  2. Construction. 150

Appendix B. Equivalent to Completeness. 157

Appendix C. Counting. 161
  1. Finite Sets. 161
  2. Countable Sets. 162
  3. Problems 165

Index 167

Introduction.

There are many, many books introducing real analysis. Feeling rather critical of each of them in some way, we wrote our own. We have thought carefully about the level and weight of each chapter and about the logical connections among the chapters, referring constantly to our familiarity with college students learning advanced mathematics for the first time.

As to the topic, we begin with a rough and ready distinction: analysis is the study of numbers and functions that are defined by successively better and better approximations; algebra is the study of the internal properties of objects defined directly. Analysis arose out of the calculus, which distinguished itself from its algebraic and geometric precursors by the use of various limits: the derivative as a limit of secant approximations, the integral as a limit of Riemann sum approximations, the infinite series as a limit of finite sums. The process of approximation was used initially to define and compute elusive quantities. As experience deepened, an insight was born, namely, that analytic representation facilitates understanding, application, and generalization. For example the series representations of the trigonometric functions allows us not only to define them more universally than by using arc length or triangles, the series elucidates their relevance to the exponential function and to the general solution of differential equations, suggesting formulas for similar functions in many other contexts.

Much of this course will resemble a repeat of calculus, but you will find that the emphasis is quite different. We are less interested in the practice of applying given techniques (e.g. differentiate these 20 functions) than in understanding why the techniques work as they do and under what circumstances they break down. We more interested in the mathematical underpinnings of an application problem than in the numerical details of such problems. Above all, we want to acquire the methodologies of approximation and estimation generally.

Students frequently have more difficulty with beginning analysis than with beginning algebra, perhaps because the axiom system in analysis is wider – there is more to keep track of. Quantifiers are stacked deeper and objects depend on each other in complicated ways. Besides reassuring you that you will get more comfortable as the course goes along, we give a rather general suggestion: When a given symbol Γ appears, ask, "What does Γ depend on?" What other objects were already on the table when Γ was defined?

You already know how to test mathematical understanding – by your ability to reproduce the salient facts in rigorous detail. You also know how the development of intuition supports your progress. As always, read actively and review regularly.

Because we are building facts up from the ground, we will give careful constructions of many common functions: the root functions, the exponential function, the trigonometric functions, the logarithm, among others. We will feel free to use these functions in examples all through the course, assuming your familiarity with them, but they will not be used in proofs until after they are constructed.

There are many exercises in the text. Almost all of them will be done in class or as homework exercises – we will make the distinction very deliberate and clear.

A note about the labeling of facts: if a proposition or theorem is preceded by an asterisk:

*Proposition:...

we expect you to remember that this fact is true – these are the basic facts of the course. If a proposition or theorem is preceded by two asterisks

**Proposition:...

then you are expected *to know a proof* of the fact. The proofs involved give the standard arguments of analysis.

A final note on terminology: if $f : A \to B$ (if f is a function from the set A to the set B), we will say that f is *onto* if for every $b \in B$, there is $a \in A$ such that $f(a) = b$. You may have used the term *surjective* for this property. We will say that f is *one to one* if $a, c \in A$ and $f(a) = f(c)$ imply that $a = c$. You may have said *injective*.

CHAPTER 1

The Real Numbers.

We will assume that the reader is familiar with elementary function composition and inverses, and with the basic properties of finite sets – these topics are covered in texts used for the Foundations of Algebra course (math 300 at Lawrence University). We will also assume you are familiar with the set $\mathbb{Z}$ of *integers*, the set $\mathbb{Q}$ of *rational numbers*, and with the algebraic and order properties of the set $\mathbb{R}$ of *real numbers*. These properties are listed in Appendix A. We will need the various intervals: *open, closed, and half-open*; these are defined in Appendix A, as well.

We assume you know the *Well-Ordering Principle* of the integers: every non-empty subset of the positive integers has a minimum element. You should also know that the use of *induction* is equivalent to that principle. Just before the list of problems in this chapter, we have provided a refresher on induction along with some exercises that will establish some of the basic properties of numbers we will use throughout the course.

Analysis lives on estimation, and estimation involves working with absolute values and inequalities. It might help to do a few exercises that stress the basics – the first couple problems at the end of the chapter. Here is what may look like a trivial example, but we will use it implicitly over and over. Suppose that $\alpha, \beta \in \mathbb{R}$ and that for all $\epsilon > 0$ we have $|\alpha - \beta| < \epsilon$. (We could say that α and β are *arbitrarily close* to each other.) We want to show that $\alpha = \beta$. Indeed, if $\alpha \neq \beta$, define $\epsilon = |\alpha - \beta|$, and then $|\alpha - \beta| < \epsilon = |\alpha - \beta|$. This contradiction shows that $\alpha = \beta$.

The previous argument may seem trivial, but it illustrates a very *analytic* way of proceeding. Often in practice we will be able to *estimate* the difference between two quantities much easier than we can compute their difference directly. It is common in analysis to prove two quantities equal by showing that they are arbitrarily close!

1. Completeness.

We begin with a fundamental axiom for the reals that is usually not discussed in a calculus course: the Completeness Property. Its use divides the advanced and more theoretical analysis from the algorithmic calculus. The Completeness Property was seen to be necessary only after the calculus had been in use for about 150 years, and so it represents a non-obvious insight. A construction of the real numbers would have to show that Completeness is true, but since we are assuming we have the real numbers already, we will assume that Completeness is true without proof. In other words, it is an *axiom*.

To introduce Completeness we need to define *boundedness*. We say that a subset S of $\mathbb{R}$ is *bounded above* if there is $b \in \mathbb{R}$ such that $x \leq b$ for all $x \in S$. This element b is called an *upper bound* for S. Because this is the first time a *for all* statement has occurred, we remind you that such a statement can be written as an implication: the statement "$x \leq b$ for all $x \in S$" is equivalent to "if $x \in S$, then $x \leq b$." Do you remember what is meant by a *vacuous implication*? That idea is used to prove that *the empty set is bounded above*.

Let S be a non-empty subset of the integers, and suppose that S is bounded above. The Well-Ordering Property implies that S has a maximum element[1] n. It is not hard to see that the integer m is an upper bound for S if and only if $m \geq n$. In other words, the number n is the *least upper bound*. The Completeness Property gives a similar statement for a non-empty subset T

---

[1]Proof: let b be an upper bound, consider the set of $1 + b - x$ where $x \in S$. Show that the $1 + b - x$ are positive and so they have a minimum. This gives S a maximum. (How?)

of the *real numbers* that is bounded above. Completeness says that there is a least (smallest) upper bound for T. Unlike in the integers, the least upper bound does not have to be the maximum of T; for instance if T is the set of negative numbers, then T does not have a maximum. (Why?) The set of upper bounds for the negative numbers is the set of non-negative numbers, and so the *least* upper bound is 0.

*COMPLETENESS PROPERTY. *Every non-empty subset of the real numbers that is bounded above has a least upper bound.*

The word *supremum* is a synonym for *least upper bound*. We use the abbreviation *sup* for the supremum.

Another example: let T be the closed interval $(-\infty, 1]$. The sup of T is the number 1, and 1 **is** an element of T. When the sup of a set is an element of that set, it is the *maximum* of the set. So, sometimes the sup of a set is an element of the set and sometimes not, when the sup arises (in a proof, in the solution of a problem) it is important to use its literal definition: *least upper bound*.

It is very easy to see that the sup of a set is unique: if $x < y$ were both sup's of the set S, then since y is the least upper bound, the number x cannot be an upper bound at all, and this is a contradiction.

Here is a very useful way to recognize a sup and to use one in a proof. The strict and non-strict inequalities in Proposition 1.1b are subtle and very important.

*PROPOSITION 1.1. *Let S be a subset of the real numbers. The real number x is the sup of S if and only if*
a) *x is an upper bound for S;*
b) *if $y \in \mathbb{R}$ and $y < x$, then there is $s \in S$ such that $y < s \leq x$.*

PROOF. Assume that x is the sup of S. Then x is an upper bound and so (a) holds. For (b), if $y < x$, then since x is the least upper bound, y cannot be an upper bound and so there is $s \in S$ with $y < s$. Since x is an upper bound, we have $s \leq x$, and (b) holds.

Now assume that (a) and (b) hold. The number x is an upper bound; is it least? If y is an upper bound for S, and if $y < x$, then (b) finds $s \in S$ with $y < s \leq x$. But $y < s$ contradicts that y is an upper bound. Thus, y is not less than x. □

Lower bounds are defined analogously to upper bounds: $x \in \mathbb{R}$ is a *lower bound* for the subset S of $\mathbb{R}$ if $x \leq s$ for all $s \in S$. A set with a lower bound is said to be *bounded below*. Analogous to least upper bound is the idea of a *greatest lower bound*. We will use the name "Completeness Property" for the following fact, even though we will give it a proof. The greatest lower bound is the *infimum*, abbreviated to *inf*.

*PROPOSITION 1.2. *Every non-empty subset of the real numbers that is bounded below has an inf (a greatest lower bound). Let S be a subset of the real numbers. The real number x is the inf of S if and only if*
a) *x is a lower bound for S;*
b) *if $y \in \mathbb{R}$ and $x < y$, then there is $s \in S$ such that $x \leq s < y$.*

PROOF. To show that the inf exists, we do the following. Let $S \subseteq \mathbb{R}$ be non-empty and bounded below, say by x. Define
$$T = \{-s \mid s \in S\}$$
and observe that T is non-empty and bounded above by $-x$. By the Completeness Property, there is a sup y for T. We claim that $-y$ is an inf for S. Indeed, let $s \in S$, then $-s \in T$ and since y is an upper bound for T, we have $-s \leq y$. It follows that $s \geq -y$, and this proves that $-y$ is a lower bound

for S. Let z be a lower bound for S, and we will show that $z \leq -y$, and this will complete the proof. Now if $s \in S$, we have $z \leq s$, and so $-z \geq -s$. This shows that $-z$ is an upper bound for T, and since y is the sup of T, we have $-z \geq y$. Thus, $z \leq -y$, as needed to prove that $-y$ is the inf of S.

The rest of the proof is left to you. □

As with sup's, inf's are unique when they exist. If the inf of a set is an element of the set, it is the *minimum* of the set.

A set with both an upper bound and a lower bound is a *bounded set*. Here is another way to say the same thing: a set S is bounded if and only if there is a closed interval $[a, b]$ such that $S \subseteq [a, b]$. One of the induction exercises showed that every non-empty finite subset of the reals has a maximum and minimum – thus, such sets are bounded. This elementary fact will be used several times in the course.

We apply Completeness to derive some basic and important properties of the integers and the rational numbers. Our first fact goes back to Archimedes; it may seem obvious.

THE ARCHIMEDEAN PROPERTY. *The set of integers is neither bounded above nor bounded below.*

PROOF. If the integers is bounded above, the Completeness Property gives it a sup x. The number $x - 1$ is not an upper bound for the integers (since x is the *least* upper bound), and so there is an integer $n > x - 1$. Then $n + 1 > x$, and this contradicts the fact that x is an upper bound. Thus, the set of integers is not bounded above.

We leave the other statement to you. □

To understand an important consequence of the Archimedean Property, we need to recall a consequence of integer Well-ordering. If S is a non-empty subset of the integers that is bounded above *by an integer*, then the set S has a maximum. We want to show that S has a maximum under the assumption that it has a *real number* upper bound (an upper bound that is not necessarily an integer). This is a direct consequence of the Archimedean Property, for let x be a real number upper bound to S. Since x is not an upper bound to the entire set of integers, there is an integer $m > x$. Then m is an *integer* upper bound for S, and so S has a maximum element.

Similarly, if a non-empty set of integers is bounded below in the reals, it has a minimum.

Here are two more consequences of the Archimedean Property.

*PROPOSITION 1.3. *Let $x \in \mathbb{R}$. There is an integer n such that $n \leq x < n+1$. If x is a positive real number, then there is a positive integer m such that $1/m < x$.*

PROOF. Given $x \in \mathbb{R}$, define S to be the set of integers m such that $m \leq x$. Because the integers is not bounded below, the set S is non-empty. Clearly S is bounded above by x, and so it has a maximum element n. Then $n \leq x$, and $n+1 \notin S$, so that $x < n+1$, as needed.

If $x > 0$, then the existence of an integer m with $1/m < x$ is the same as the existence of m such that $1/x < m$. □

2. Rational and Irrational Density.

It is a subtle but important fact that between every pair of distinct real numbers, there are infinitely many rational numbers. We say that the rational numbers are *dense* in the set of real numbers.

**RATIONAL DENSITY. *Let $x, y \in \mathbb{R}$ with $x < y$. Then there is a rational number r such that $x < r < y$.*

PROOF. The number $y - x$ is positive, and so Proposition 1.3 finds an integer n with $1/n < y - x$. The set S of integers m with $m > n \cdot x$ is bounded below, and so it has a minimum element k. We claim that the rational number k/n is properly between x and y.

The definition of k shows that $k > n \cdot x$, so that $x < k/n$. Since k is the minimum, we must have $k - 1 \leq n \cdot x$; in other words $k \leq n \cdot x + 1$. Use that $1/n < y - x$ to see that

$$\frac{k}{n} \leq \frac{n \cdot x + 1}{n} = x + \frac{1}{n} < x + y - x = y$$

and we have $k/n < y$. □

The real numbers that are not rational are *irrational*. You probably know that $\sqrt{2}$ is irrational. We mentioned before that in formal terms we will **not** assume the existence of square roots or other roots until we construct them. Once $\sqrt{2}$ is shown to exist, the following proof will stand – it says that the irrational numbers are dense.

*PROPOSITION 1.4. *Let $x, y \in \mathbb{R}$ with $x < y$. Then there is an irrational number r such that $x < r < y$.*

PROOF. Since $\sqrt{2} > 0$, we have $x/\sqrt{2} < y/\sqrt{2}$. By Rational Density, there is $r \in \mathbb{Q}$ such that

$$\frac{x}{\sqrt{2}} < r < \frac{y}{\sqrt{2}}$$

We need r to be non-zero. If $r = 0$, then use Rational Density again to get a rational number r' between $x/\sqrt{2}$ and 0, and then $r' \neq 0$. In this case, replace r by r'. In any case, we have a non-zero rational number between $x/\sqrt{2}$ and $y/\sqrt{2}$. Multiplying by $\sqrt{2}$, we obtain

$$x < r \cdot \sqrt{2} < y$$

The number $r \cdot \sqrt{2}$ is irrational, for if it were rational, then since $r \neq 0$, we have that $1/r$ is rational, and the product of rational numbers is rational:

$$\frac{1}{r} \cdot r \cdot \sqrt{2} = \sqrt{2}$$

This contradicts that $\sqrt{2}$ is irrational. □

Review of Induction

Induction is a property of the set of positive integers. It says that if T is a subset of the set of positive integers, then T is the entire set of positive integers, provided that the following two statements are true.

(1) $1 \in T$
(2) $n \in T$ implies that $(n+1) \in T$

This certainly makes intuitive sense; it is a non-trivial fact that it does not follow from the arithmetical properties of the integers; we regard it as an axiom. Here is a standard induction proof the way it would be written out.

Example. Every non-empty finite subset of the real numbers has a maximum.

Solution. Induction on the number of elements in the finite set F. If F has exactly one element: $F = \{x\}$, then x is obviously the maximum of F. Let n be a positive integer, and assume that each set with n elements has a maximum. Let F have $n+1$ elements. Choose $x \in F$, and define $F' = F \setminus \{x\}$, so that F' has n elements. Then F' has a maximum c. If $c > x$, then c is the maximum of F. If $x > c$, then x is also greater than every element of F', and so x is the maximum of F. We conclude that each set with $n+1$ elements has a maximum. ■

We have written the proof in a style typical of induction proofs. Here is more formal detail that shows how the literal induction axiom is being used.

Formal Solution Let T be the set of positive integers n such that if $F \subset \mathbb{R}$ and F has n elements, then F has a maximum. We prove that $1 \in T$ and that $n \in T$ implies $(n+1) \in T$. It will follow that T is the entire set of positive integers. Thus, if F is a non-empty finite set of real numbers with n elements, then $n \in T$, and so F will have a maximum element.

The proof that $1 \in T$ is the proof above that if F has 1 real number element, then that one element is the maximum. The proof that $n \in T$ implies $(n+1) \in T$ occurred above when we assumed that each set with n elements has a maximum, and we considered a set with $n+1$ elements. ∎

Here are some elementary properties of the real numbers that will be used fairly often. Each one can be proved by induction.

- ★(I1) Every non-empty finite set of real numbers has a minimum.
- (I2) Every non-empty finite set of real numbers F can be put into order. In other words, we can subscript the elements of F, so that those elements are $f_1 < f_2 < \cdots < f_n$
- ★(I3) If n is a positive integer, and $x, y \in \mathbb{R}$ with $0 \leq x < y$, then $x^n < y^n$.
- ★(I4) Let $\alpha \in \mathbb{R}$ with $\alpha \geq 2$. Then $\alpha^n > n$ for all positive integers n.

3. Problems

1. (Proving an inequality by estimation.) Suppose that $\alpha, \beta \in \mathbb{R}$ and that $\alpha < \beta + \epsilon$ for all $\epsilon > 0$. Then $\alpha \leq \beta$.

2. Recall the definition of the absolute value: for $x \in \mathbb{R}$, we have $|x| = x$ if $x \geq 0$ and $|x| = -x$ if $x < 0$. Use this definition to prove the *Triangle Inequalities*: if $a, b \in \mathbb{R}$, then $|a+b| \leq |a| + |b|$ and $|a-b| \geq |a| - |b|$.

3. Use the definition of the absolute value to prove that if $x, y \in \mathbb{R}$, then $|x| \leq y$ if and only if $-y \leq x \leq y$.

4. Let S be the set of $x \in \mathbb{R}$ such that $x^4 - 7 \cdot x^3 - 30 \leq 5$. Show that S is bounded above. (Hint: prove the contrapositive.)

5. Let $S \subseteq \mathbb{R}$ and suppose that S is bounded above. Let B be the set of upper bounds of S and suppose that B has no lower bound. What do you conclude about S?

6. Let S be the set of $x \in \mathbb{R}$ such that $x^2 \cdot \exp(-3x) \geq 0.06$. Show that S is bounded above and that S is not bounded below.[2]

7. Let $f(x)$ be a polynomial, and let a, b be numbers with $a \leq b$. Let S be the set of $f(x)$ for $a \leq x \leq b$. Show that S is bounded above. (Hint: show that there is a number $c > 0$ such that $|x| \leq c$ and use c to estimate $f(x)$. It is ok for the upper bound to depend on a, b, c, but not on x.)

8. Let $x, y \in \mathbb{R}$ with $x < y$. Then y is sup of the open interval (x, y), the closed interval $[x, y]$ and each of the half-open intervals $(x, y]$ and $[x, y)$.

9. Let S, T be subsets of the reals with sups α, β, respectively. Show that $\alpha + \beta$ is the sup of the set of all $x + y$ where $x \in S$ and $y \in T$.

10. For subsets S, T of the reals, define $S * T$ to be the set of all $x \cdot y$ where $x \in S$ and $y \in T$. Find subsets S, T of the reals each of which is bounded above, but with $S * T$ is not bounded above. *Think of one infinite* ★

11. Complete the proof of Proposition 1.2: show that x is the inf if and only if (a) and (b) hold.

12. Let S be the set of $x \in \mathbb{R}$ such that $x^2 - \sin(x) \leq 5$. Prove that S is bounded.

13. (Part of the Archimedean Property) Prove that the set of integers is not bounded below.

14. (An alternative form of the Archimedean Property) Let a, b be positive real numbers. Then there is an integer n with $n \cdot a > b$.

15. (A stronger form of Rational Density) Let $x, y \in \mathbb{R}$ with $x < y$. Then there are infinitely many rational numbers in the open interval (x, y). (Hint: assume there are finitely many, so that there is a maximum.)

---

[2] Recall the standard notation $\exp(x) = e^x$. Since this problem uses e^x in an example, you should feel free to use any of the properties of this function that you know from calculus.

16. Prove the following theorem of Dedekind:[3] Suppose that $\mathbb{R}$ is the disjoint union of non-empty sets L and R, and suppose that if $x \in L$ and $y \in R$, then $x < y$. Then there is $\alpha \in \mathbb{R}$ such that if $x \in L$, then $x \leq \alpha$, and if $y \in R$, then $\alpha \leq y$.

17. Take Dedekind's Theorem (in the previous problem) as an *axiom*. Prove the Completeness Property. (Hint: Let S be non-empty and bounded above. Let R be the set of upper bounds of a set S, and let $L = \mathbb{R} \setminus R$. Be careful about the case that S has a maximum.)

---

[3] See p.30 of *A Course in Pure Mathematics*, by G.H. Hardy, 10th ed., Cambridge University Press, 1958.

CHAPTER 2

Sequences.

1. Definitions and Examples.

A *sequence* is an infinite list of numbers. It is customary to use subscripts to denote the ordering in the list: $a_1, a_2, a_3, \ldots$. We started with subscript 1, but we can start with any integer, so that $b_{-2}, b_{-1}, b_0, \ldots$ would also denote a sequence.

In formal terms, a sequence is a function from the set of subscripts to the set of real numbers. If the subscripts start at p, say, then we would have a function

$$f : \{k \in \mathbb{Z} \mid k \geq p\} \to \mathbb{R}$$

The subscript notation f_k is just an alternative way of expressing the function notation $f(k)$.

It is common to refer to a sequence by a symbol that looks like a single value of the function: to say, "Let a_k be a sequence." What we really mean (formally) is that a is a sequence; the symbol a_k is "the function a evaluated at k." Writing a_k for the sequence is like writing the function f as $f(x)$, including a name for the variable.

You will sometimes see the notation $\{a_n\}$ for the sequence a_n. We will not use this notation, since the set brackets make a sequence look like a set, and this obscures the fact that the sequence values are *listed*. For instance, numbers in a *set* do not *repeat*, numbers in a sequence can repeat as often as they wish.

Some texts simply define a sequence as a function from the positive integers into the reals: $a_1, a_2, \ldots$. This is *equivalent* to our definition, since we can rewrite a list with subscripts that start with 1, but it will be convenient to have the more general definition. The texts that give the definition starting with a_1 almost immediately consider sequences that start a_0, and so nothing is really gained by being more specific. In most examples of sequences, the domain of subscripts is obvious from context, and so we won't call much attention to it.

You have seen many examples of sequences already, such as the *harmonic sequence* $1/n$ for $n \geq 1$. Many sequences arise in practice *recursively*. For instance, the *factorial sequence* can be defined like this:

$$0! = 1 \quad \text{and} \quad (n+1)! = (n+1) \cdot n! \quad \text{for} \quad n = 0, 1, 2, \ldots$$

Hopefully, the study of rigorous mathematics has taught you to be suspicious of the "..." notation. In this context, we mean that n is a non-negative integer. Taking successive values $n = 0$, then $n = 1$, then $n = 2$, and so on, produces successive values of the factorial sequence.

For $r \in \mathbb{R}$ with $r \neq 0$, it is useful to define the simple exponential r^n recursively as well.

$$r^0 = 1 \quad \text{and} \quad r^{n+1} = r \cdot r^n \quad \text{for} \quad n = 0, 1, 2, \ldots$$

Induction and recursion are closely related: induction can be used to prove that a recursive definition actually defines a sequence. It will be useful to us that induction can often be used to establish properties of a recursively defined sequence – we will see examples momentarily.

Related to the factorials, for each real number α there is a sequence called the *binomial sequence*. It is defined informally by these equations:

$$\binom{\alpha}{0} = 1 \quad \text{and} \quad \binom{\alpha}{n} = \frac{\alpha \cdot (\alpha - 1) \cdots (\alpha - n + 1)}{n!} \quad \text{for} \quad n = 1, 2, 3, \ldots$$

If we were to use sequence notation, we would write $\binom{\alpha}{n} = b_n$, but in the notation b_n, we would need to remember that a specific number α is being used to define the sequence values.

Here is a more formal definition of the binomial sequence – using recursion. We hope to show that this more mysterious definition is actually more useful!

$$\binom{\alpha}{0} = 1 \quad \text{and} \quad \binom{\alpha}{n+1} = \frac{\alpha - n}{n+1} \cdot \binom{\alpha}{n} \quad \text{for} \quad n = 0, 1, 2, \ldots$$

You might know that when α is a non-negative integer, the numbers $\binom{\alpha}{n}$ are 0 when $n > \alpha$. The non-zero values for $0 \leq n \leq \alpha$ are the numbers that occur in *Pascal's Triangle*; they count the number of ways to choose a set of n objects from a set of α distinct objects. The binomial sequence also occurs in the binomial theorem.[1]

Here is an ancient sequence used, as we will see, to approximate square roots. Given a real number $\alpha > 1$, define the *Babylonian sequence for the square root*

$$a_0 = \alpha, \quad a_{n+1} = \frac{1}{2} \cdot \left(a_n + \frac{\alpha}{a_n} \right) \quad \text{for} \quad n \geq 0$$

As with the binomial sequence, the notation a_n assumes we have fixed the number α that creates the sequence. It is easy to see that the a_n are positive. Later, we will show that this sequence converges to $\sqrt{\alpha}$; this will prove that the square root exists. You can experiment with this sequence (we recommend you try $\alpha = 2$, for instance) to see evidence that it approximates the square root.

2. Convergence.

When we say that a sequence a_k has a property *eventually* we mean that there is a real number N such that if $k \geq N$, then a_k has the property. For instance, the sequence $a_k = 1/k^2$ for $k \geq 0$ is *eventually less than* $1/10$, for

---

[1] We will discuss a very general form of the binomial theorem in Chapter 8 – see p.142.

if $k \geq 4$, then $a_k < 1/10$. (Prove this!) The sequence $b_k = k^2 - \sqrt{5} \cdot k$ is *eventually positive*, since if $k \geq 3$ we have $b_k > 0$. (Prove!)

Another example: when α is a non-negative integer, the sequence $\binom{\alpha}{n}$ is eventually zero. (We already mentioned this property, without using the word *eventually*.) Indeed, the recursive formula shows that
$$\binom{\alpha}{\alpha+1} = \frac{\alpha - \alpha}{\alpha + 1} \cdot \binom{\alpha}{\alpha} = 0$$
The recursive formula then shows that $\binom{\alpha}{\alpha+2} = 0$, as well, and then $\binom{\alpha}{\alpha+3} = 0$, and so on. (Induction!)

The existence of a real number N such that $n \geq N$ implies some property of a_n is equivalent to the existence of an *integer* N such that $n \geq N$ implies the property. Indeed, given the real number N, the Archimedean Property of the reals finds an integer $M \geq N$, and then if $n \geq M$, then $n \geq N$, and a_n has the required property. In other words, if we want to, we can make the definition of *eventually* to require the existence of an *integer* N such that $n \geq N$ implies what we want. It is sometimes convenient to have the N be an integer, and sometimes it doesn't matter.

You have seen the idea of the limit of a sequence a_k. We want to make that idea precise – this is first major example of analytic technique. The notion of the limit involves letting k go to infinity: $k \to \infty$. Here is a precise formulation: Given a sequence a_n and a real number A, we say "$a_k \to A$ as $k \to \infty$" to mean this: if $\epsilon > 0$ is given, then there is a real number N such that if $k \geq N$, then $|a_k - A| < \epsilon$. To repeat: given $\epsilon > 0$, we have $|a_k - A| < \epsilon$ *eventually*.

Here is a somewhat informal but useful restatement: $a_k \to A$ as $k \to \infty$ means that a_k is eventually arbitrarily close to A. As we work with the limit, we will explore all the logic cleverly hidden in this definition.

2. CONVERGENCE.

You have seen this notation for the limit: $\lim_{k\to\infty} a_k = A$. We also say that the sequence *converges* to A. If we don't want to mention the number A, we say that the sequence *converges*. A sequence which doesn't converge is said to *diverge*.

Some simple and familiar examples: consider the sequence $1/k$ for $k \geq 1$, and we'll show that $1/k \to 0$ as $k \to \infty$. To prove this, we need to be given an arbitrary positive number ϵ, and we need to show that $|1/k - 0| < \epsilon$ eventually. This inequality is $1/k < \epsilon$, which is equivalent to $1/\epsilon < k$. To straighten out the logic: **If $k > 1/\epsilon$, then** $|1/k - 0| < \epsilon$. Or we can say: $1/k < \epsilon$ eventually. This proves that $1/k \to 0$ as $k \to \infty$.

Next we show that $(-1)^k$ for $k \geq 0$ cannot have a limit. For if A were a limit, take $\epsilon = 1$ and we would have $|(-1)^k - A| < 1$ eventually. In particular, there would be k such that

$$|(-1)^k - A| < 1 \quad \text{and} \quad |(-1)^{k+1} - A| < 1$$

Check the following calculation carefully, noting that $(-1)^k - (-1)^{k+1}$ is either 2 or -2.

$$2 = |(-1)^k - (-1)^{k+1}| = |(-1)^k - A + A - (-1)^{k+1}|$$
$$\leq |(-1)^k - A| + |A - (-1)^{k+1}| < 1 + 1 = 2$$

so that $2 < 2$, a contradiction.

When a limit exists, it is unique.

*PROPOSITION 2.1. *Let a_k be a sequence, and suppose that $a_k \to A$ as $k \to \infty$ and that $a_k \to B$ as $k \to \infty$. Then $A = B$.*

PROOF. Let $\epsilon > 0$, and there is a real number a such that if $k \geq a$, then $|a_k - A| < \epsilon$. There is also a real number b such that if $k \geq b$, then $|a_k - B| < \epsilon$. By the Archimedean Property, there is a positive integer k greater than a and

greater than b. For this particular value of k, we have both $|a_k - A| < \epsilon$ and $|b_k - B| < \epsilon$, and so we can compute
$$|A - B| = |A - a_k + a_k - B| \leq |A - a_k| + |a_k - B| < \epsilon + \epsilon = 2\epsilon$$
The number 2ϵ is an arbitrary positive number, and we see that $A = B$. □

One consequence of Proposition 2.1 involves the notation $\lim_{k \to \infty} a_k$. First, if a_k has no limit, this notation is meaningless – a dangerous status for mathematical notation; in fact, we prefer to avoid this notation whenever we can. If, on the other hand, a_k has a limit, then $\lim_{k \to \infty} a_k$ can be taken to be the unique limit and used as a number. Even here we have to be careful, as we will point out in proving the algebraic properties of the limit.

Another elementary fact that is very useful.

PROPOSITION 2.2. *Let a_k be a sequence, and suppose that $a_k \to A$ as $k \to \infty$. Let y a real number with $y < A$. Then $y < a_k$ eventually. Similarly, if $A < y$, then $a_k < y$ eventually.*

PROOF. Suppose that $A < y$, and let $\epsilon = y - A$, a positive number. Since $a_k \to A$ as $k \to \infty$, there is a positive integer n such that if $k \geq n$, then $|a_k - A| < \epsilon$. For $k \geq n$, compute
$$a_k - A \leq |a_k - A| < \epsilon = y - A$$
and we conclude that $a_k - A < y - A$, which leads to $a_k < y$. The proof for the case $y < A$ is similar. □

The contrapositive of Proposition 2.2: if $y \in \mathbb{R}$ and $a_k \geq y$ for all k and if $a_k \to A$, then $A \geq y$. We might say, "Limits preserve non-strict inequalities." We should note that limit do not necessarily preserve strict inequalities, for $1/k > 0$ for $k = 1, 2, \ldots$, but the limit $1/k \to 0$ is not positive.

Proposition 2.2 has an analogue with the inequalities reversed. For instance, if $a_k \leq y$ for all k, and if $a_k \to A$, then $A \leq y$.

The number B is an *upper bound* for the sequence a_k if $a_k \leq B$ for all k. The phrase "for all k" tells us that B is independent of k. As with upper bounds of sets, upper bounds for sequences are not unique. If a sequence has an upper bound it is *bounded above*. Similarly, a sequence a_k is *bounded below* if it has a *lower bound* B such that $B \leq a_k$ for all k. A sequence is *bounded* if it has both an upper and a lower bound. Equivalently, a sequence a_k is bounded, if there is a number C such that $|a_k| \leq C$ for all k.

Convergent sequences are bounded.

**PROPOSITION 2.3. *Let a_k be a sequence, and suppose that $a_k \to A$ as $k \to \infty$. Then there is a real number B such that $|a_k| \leq B$ for all k.* (bounded above)

PROOF. Suppose that a_k is defined for $k \geq L$. Taking 1 as "ϵ," there is a positive integer n such that if $k \geq n$, then $|a_k - A| < 1$. We can assume that $n \geq L$, since a_k is not defined when $k < L$. Notice that the set

$$\{|A| + 1, |a_L|, |a_{L+1}|, \ldots, |a_n|\}$$

is finite. Therefore, it has a maximum B. We claim that $|a_k| \leq B$ for all k. Indeed, if $L \leq k \leq n$, then by the definition of B, we have $|a_k| \leq B$. For $k > n$, we have $|a_k - A| < 1$, and so

$$|a_k| = |a_k - A + A| \leq |a_k - A| + |A| < 1 + |A| \leq B$$

as needed. □

The sequence $(-1)^k$ is bounded but does not converge, so we need to be careful to apply Proposition 2.3 as it is stated. Sometimes it is useful to have the strict inequality $|a_k| < B$ for all k rather than $|a_k| \leq B$. It is obvious that if we take B as in the conclusion of Proposition 2.3, then we can use $B+1$ to get the strict inequality $|a_k| < B + 1$.

Our next result is the familiar limit algebra. Most of this is easy, but we need to be aware of a couple of issues. First, given sequences a_k and b_k, we want to consider the sequences $a_k + b_k$ and $a_k \cdot b_k$ and a_k/b_k. Obviously, when we write $a_k + b_k$, we are using the same k for both a_k and b_k and so we need both a_k and b_k to be defined. If a_k is defined for $k \geq 0$, say, and b_k for $k \geq 10$, then $a_1 + b_1$ is not defined. However, once $k \geq 10$, both a_k and b_k are defined and we can write a_k+b_k. Since limits have to do with what happens eventually, we will not worry about losing $a_0, a_1, \ldots, a_9$ when we speak of $a_k + b_k$. More formally: let a_k be defined for $k \geq L$, and let b_k be defined for $k \geq M$. Let N be the maximum of M, L, and notice that a_k, b_k are both defined for $k \geq N$. The sequences $a_k + b_k$ and $a_k \cdot b_k$ will be defined for $k \geq N$. (We might have said that $a_k + b_k$ is *eventually* defined!)

To make sense of a_k/b_k, we need that the denominator is not 0 – "eventually not 0" will do.

PROPOSITION 2.4. *Let a_k, b_k be sequences, and suppose that $a_k \to A$ as $k \to \infty$, and that $b_k \to B$ as $k \to \infty$. Then we have the following.*
a) $(a_k + b_k) \to (A + B)$ *as* $k \to \infty$.
**b)* $(a_k \cdot b_k) \to (A \cdot B)$ *as* $k \to \infty$.
c) *If $B \neq 0$, then b_k is eventually not zero, and $(a_k/b_k) \to (A/B)$ as $k \to \infty$.*

PROOF. (a): Let $\epsilon > 0$. There is a positive integer n such that if $k \geq n$, then $|a_k - A| < \epsilon$, and there is a positive integer m such that if $k \geq m$, then $|b_k - B| < \epsilon$. Let p be the maximum of $\{n, m\}$ and then $a_k + b_k$ is defined for $k \geq p$, and we also have $|a_k - A| < \epsilon$ and $|b_k - B| < \epsilon$. Then

$$|(a_k + b_k) - (A + B)| = |(a_k - A) + (b_k - B)|$$
$$\leq |a_k - A| + |b_k - B| < \epsilon + \epsilon = 2\epsilon$$

We have shown that $|(a_k+b_k)-(A+B)|$ is eventually less than 2ϵ. A technical point: the definition of the limit says we need to show that the sequence minus the limit is eventually less than ϵ. Is 2ϵ too big? The answer is, no. The ϵ

of the definition of the limit is a placeholder for an arbitrary positive number. The number 2ϵ that showed up in our argument is also an arbitrary positive number. Thus, we have shown that $|(a_k+b_k)-(A+B)|$ is eventually arbitrarily small, and this establishes that $(a_k + b_k) \to (A + B)$ as $k \to \infty$.

The fussing we just did (maximum of n, m, arguing over 2ϵ) is necessary to the other statements, but we will be more cavalier and leave some of these details to you, giving our arguments at a more customary level.

For (b), we use an old algebra trick.
$$(a_k \cdot b_k) - A \cdot B = (a_k - A) \cdot b_k + A \cdot (b_k - B)$$

Proposition 2.3 finds a number C such that $|b_k| < C$ for all k. Given $\epsilon > 0$, we have $|a_k - A| < \epsilon$ eventually, and we have $|b_k - B| < \epsilon$ eventually. Thus, we eventually have

$$\begin{aligned}\left|(a_k \cdot b_k) - A \cdot B\right| &= \left|(a_k - A) \cdot b_k + A \cdot (b_k - B)\right| \\ &\leq \left|(a_k - A) \cdot b_k\right| + \left|A \cdot (b_k - B)\right| \\ &\leq \left|a_k - A\right| \cdot \left|b_k\right| + \left|A\right| \cdot \left|b_k - B\right| \\ &< \epsilon \cdot C + |A| \cdot \epsilon\end{aligned}$$

The number $\epsilon \cdot (C + |A|)$ is arbitrary positive, and so $(a_k \cdot b_k) \to (A \cdot B)$.

For (c), we will work first just with b_k. We'll show that $1/b_k \to 1/B$. Using $|B|/2$ as an "ϵ", we see that $|b_k - B| < |B|/2$ eventually. For these k, use the Triangle Inequality to compute

$$|b_k| = |B - (B - b_k)| \geq |B| - |B - b_k| > |B| - |B|/2 = |B|/2$$

This proves that b_k is eventually non-zero. We will consider the k such that $|b_k| > |B|/2$. Given $\epsilon > 0$, we have eventually that $|b_k - B| < \epsilon$, and we compute

$$\left|\frac{1}{b_k} - \frac{1}{B}\right| = \left|\frac{B - b_k}{B \cdot b_k}\right| = \frac{|B - b_k|}{|B| \cdot |b_k|} < \frac{\epsilon}{|B| \cdot |B|/2}$$

This number is arbitrarily small. This proves that $1/b_k \to 1/B$ as $k \to \infty$. Now (b) proves that $a_k/b_k \to A/B$, which is (c). □

Proposition 2.4b includes the following case: if $a_k \to A$ as $k \to \infty$, and if β is a constant, then $\beta \cdot a_k \to \beta \cdot A$ as $k \to \infty$. Indeed, we just define $b_k = \beta$, so that $b_k \to \beta$.

The limit algebra of Proposition 2.4 covers the most common cases of limits. Here is a representative calculation from calculus.
$$\lim_{n \to \infty} \frac{3n+2}{4n-101}$$
Since we know the limit of $1/n$, we try to get that expression to occur:
$$\frac{3n+2}{4n-101} = \frac{3+2/n}{4-101/n}$$
Look at the $2/n$ term. Since $2 \to 2$, Proposition 2.4b shows that $2/n = 2 \cdot (1/n) \to 2 \cdot 0 = 0$. Similarly, the other parts of the fraction go to the expected limits, and Proposition 2.4 puts these limits together:
$$\frac{3+2/n}{4-101/n} \to \frac{3+0}{4-0} = \frac{3}{4}$$
The idea is to obtain the limit *without* using the limit definition directly. This sort of algebraic calculation can be learned without really understanding the definition of the limit – that's what almost all students do in calculus. In this course, we will emphasize the logic that flows out of the literal definition, making liberal use of the limit algebra whenever we can.

Here is another common situation. Notice that the sequence b_k in the following does not have to have a limit!

*Proposition 2.5. Let a_k, b_k be sequences, and suppose that $a_k \to 0$ as $k \to \infty$, and assume that b_k is bounded. Then $(a_k \cdot b_k) \to 0$ as $k \to \infty$.

Proof. Let $|b_k| \leq B$ for all k. Let $\epsilon > 0$ and then eventually, $|a_k| < \epsilon$. It follows that $|a_k \cdot b_k| \leq B \cdot \epsilon$. □

For instance $\frac{1}{n} \cdot \sin(n) \to 0$ as $n \to \infty$.

Next we come to what is called the "squeeze" law. It has always seemed obvious to me, but it gets a proof in every analysis text.

PROPOSITION 2.6. *Let a_k, b_k, c_k be sequences and suppose that $a_k \leq b_k \leq c_k$ eventually. Suppose that $a_k \to A$ and $c_k \to A$ as $k \to \infty$. Then $b_k \to A$ as $k \to \infty$.*

PROOF. If $a_k \leq b_k \leq c_k$, then
$$b_k - A \leq c_k - A \leq |c_k - A| \quad \text{and} \quad A - b_k \leq A - a_k \leq |A - a_k|$$
If k is large enough, then both $|c_k - A|$ and $|A - a_k|$ are less than a given $\epsilon > 0$. And we then have $b_k - A < \epsilon$ and $A - b_k < \epsilon$, so that $|b_k - A| < \epsilon$. □

Finally we come back to the issue of subscripts. Given a sequence a_k, it is sometimes interesting to consider the sequence a_{k+1}. Technically, the definition goes like this: If a_k has $k \geq L$, then $b_k = a_{k+1}$ is defined for $k \geq L$, as well. We often think of b_k as the same *list* as a_k except that the first term in the list has been omitted. The main point is this: whatever happens *eventually* to a_k happens also *eventually* to b_k. For instance, if $a_k \to A$ as $k \to \infty$, then also $b_k \to A$. This is usually expressed by saying that $a_{k+1} \to A$. We will use this idea in the next section.

3. Monotone Convergence.

If a sequence a_k satisfies $a_k \geq a_{k+1}$ for each k, then we say that a_k is *decreasing*. This term might seem a little inaccurate, since decreasing might seem to mean that $a_k > a_{k+1}$. However, it is customary to allow equality. To mean that $a_k > a_{k+1}$, we say that the sequence is *strictly decreasing*. The sequence a_k is *eventually decreasing* if there is N such that $k \geq N$ implies that $a_k \geq a_{k+1}$.

Similarly we say that a_k is *increasing* if $a_k \leq a_{k+1}$ for each k; it is eventually increasing if there is N such that $a_k \leq a_{k+1}$ for $k \geq N$. The sequence is *strictly increasing* if $a_k < a_{k+1}$.

The deeper theorems about sequence limits give us the *existence* of a limit without knowing exactly what the limit is. The following theorem of this sort is extremely fundamental. It is, in fact, equivalent to the Completeness Property of the real numbers. We have regarded Completeness as a axiom, and we will give a proof of this theorem. In an appendix, we will turn the tables and show how Completeness can be derived from the Monotone Convergence Theorem.

**MONOTONE CONVERGENCE THEOREM. *Let a_k be a sequence. If a_k is eventually increasing and bounded above, then it converges. If a_k is eventually decreasing and bounded below, then it converges.*

PROOF. We will do the case where a_k is eventually increasing, leaving the decreasing case to class. Suppose that n is a positive integer such that if $k \geq n$, then $a_k \leq a_{k+1}$. It is easy to see that

$$a_n \leq a_{n+1} \leq a_{n+2} \leq \cdots$$

We are assuming that the sequence is bounded above, and so the set of sequence values a_k for $k \geq n$ is a non-empty set that is bounded above. Completeness gives this set a sup[2] A. We will prove that $a_k \to A$ as $k \to \infty$.

Given $\epsilon > 0$, the number $A - \epsilon$ is not an upper bound for the a_k with $k \geq n$. Thus, there is $k \geq n$ such that $A - \epsilon < a_k$. Since the sequence is increasing at this point and since A is an upper bound, we have

$$A - \epsilon < a_k \leq a_{k+1} \leq \cdots \leq A$$

We see that if $j \geq k$, then $|A - a_j| < \epsilon$. This proves that $a_k \to A$. □

---

[2] Note that A is the sup of $a_n, a_{n+1}, \ldots$, not necessarily the sup of the *entire* sequence.

Here are some uses of Monotone Convergence, beginning with some limits that are familiar. In each case notice how recursion is used.

When r is a real number with $|r| < 1$, we have $r^n \to 0$ as $n \to \infty$. We'll prove that $|r|^n \to 0$, and an exercise then shows that $r^n \to 0$. Since $|r| < 1$, we see that $|r|^n$ is a decreasing sequence. Furthermore, $|r|^n > 0$, so this sequence is bounded below. Monotone Convergence says it must have a limit A. Now use the fact that the sequence $|r|^{n+1}$ has the same limit as the sequence $|r|^n$. Recursion $|r|^{n+1} = |r| \cdot |r|^n$, combined with the limit algebra, gives this

$$A = |r| \cdot A \quad \text{so that} \quad A \cdot (1 - |r|) = 0$$

Since $|r| < 1$, this implies that $A = 0$.

The foregoing gives an expected result. Notice, however, that the existence of the limit A was established before we proved that $A = 0$. This methodology occurs frequently in advanced mathematics: I have the existence of a number before I know what it is.

Another one: when r is a real number with $r > 1$, we will show that r^n is not bounded above.[3] For if the sequence r^n is bounded above, Monotone Convergence gives it a limit A that is the sup of the numbers r^n. In particular, $A \geq r$, and so $A \neq 0$. Then $r^{n+1} \to A$ as well, and so

$$r^{n+1} = r \cdot r^n \to r \cdot A$$

shows that $A = r \cdot A$. Since $A \neq 0$, this implies that $r = 1$, which is not the case. This proves that r^n is not bounded above.

A slightly more profound result. Let $\alpha > 1$ be given and recall the Babylonian sequence a_n described above. We will show that $a_{n+1} < a_n$ and $\alpha < a_n^2$ for all $n \geq 0$. We start with $a_0 = \alpha$, so that $a_0^2 = \alpha^2 > \alpha$, since $\alpha > 1$.

---

[3] You probably know that we say $r^n \to \infty$ here. We will introduce limits to infinity later.

For an induction argument, assume that $a_n^2 > \alpha$ for some n, and compute

$$a_{n+1} = \frac{1}{2} \cdot \left(a_n + \frac{\alpha}{a_n}\right) < \frac{1}{2} \cdot \left(a_n + \frac{a_n^2}{a_n}\right) = \frac{1}{2} \cdot (a_n + a_n) = a_n$$

Next compute that

$$(a_n^2 - \alpha)^2 > 0 \qquad \text{square out}$$
$$a_n^4 - 2 \cdot \alpha \cdot a_n^2 + \alpha^2 > 0 \qquad \text{divide by } a_n^2$$
$$a_n^2 - 2 \cdot \alpha + \frac{\alpha^2}{a_n^2} > 0 \qquad \text{add } 4 \cdot \alpha$$
$$a_n^2 + 2 \cdot \alpha + \frac{\alpha^2}{a_n^2} > 4 \cdot \alpha \qquad \text{note square on the left}$$
$$\left[a_n + \frac{\alpha}{a_n}\right]^2 > 4 \cdot \alpha$$
$$\frac{1}{4} \cdot \left[a_n + \frac{\alpha}{a_n}\right]^2 > \alpha \quad \text{so that} \quad a_{n+1}^2 > \alpha$$

Starting with $a_0^2 > \alpha$, we get $a_0 > a_1$ and $a_1^2 > \alpha$. Then we get $a_1 > a_2$ and $a_2^2 > \alpha$. And so on.

The sequence a_n is decreasing; and it is bounded below by 1, since $1 < \alpha < a_n^2$ for all n. By the Monotone Convergence Theorem, the sequence converges, say to A, and we know that $A \geq 1$. Also, $a_{n+1} \to A$ as well, and the limit algebra shows that

$$a_{n+1} = \frac{1}{2} \cdot \left(a_n + \frac{\alpha}{a_n}\right) \quad \text{implies that}$$
$$A = \frac{1}{2} \cdot \left(A + \frac{\alpha}{A}\right)$$

Manipulating this last equation yields $A^2 = \alpha$. Since $A \geq 1$, we see that we have constructed a non-negative number whose square is equal to α; we'll write $A = \sqrt{\alpha}$ as usual.

3. MONOTONE CONVERGENCE.

Since we assumed that $\alpha > 1$, we have the existence of square roots of numbers greater than 1. It is an exercise to get square roots of the other non-negative numbers.

There is a similar construction of cube roots, fourth roots, and so on. What we are really doing here is using *Newton's Method*, which you may or may not remember from calculus. We won't pursue this technique, but you are invited to look it up for yourself.

Now that we have square roots, we can define fourth roots: $x^{1/4} = \sqrt{\sqrt{x}}$. We can use this idea to introduce an intriguing sequence. Given the number $\alpha > 1$, define the sequence a_n where $a_0 = 1$ and $a_{n+1} = (\alpha \cdot a_n)^{1/4}$ for $n \geq 0$. In an exercise, you will investigate this sequence.   #15

We close this chapter with one more fact. Several constructions later in the course will involve sequences that approach each other from two directions; here is the set-up and the conclusion we want in that setting.

*PROPOSITION 2.7. *Suppose we have sequences a_k and b_k such that*

$$a_0 \leq a_1 \leq \cdots \leq a_n < b_n \leq \cdots \leq b_1 \leq b_0 \quad \text{for each} \quad n$$

Suppose also that $(b_n - a_n) \to 0$ as $n \to \infty$. Then there is a real number A such that

a) $a_n \to A$ and $b_n \to A$ as $n \to \infty$;
b) $A \in [a_n, b_n]$ for all n;
c) If $\delta > 0$, then there is n such that $[a_n, b_n] \subseteq (A - \delta, A + \delta)$.

PROOF. The sequence a_k is increasing and bounded above by each b_n, and so it converges to its sup A, by the Monotone Convergence Theorem. Since A is an upper bound for a_k, we have $a_n \leq A$ for all n. Since A is least and each b_n is an upper bound, $A \leq b_n$ for all n. Thus, $A \in [a_n, b_n]$, which proves (b).

Since $(b_n - a_n) \to 0$ and $a_n \to A$, Proposition 2.4a shows that

$$b_n = (b_n - a_n) + a_n \to 0 + A = A \quad \text{as} \quad n \to \infty$$

We have proved (a).

For (c), let $\delta > 0$. Since $(b_n - a_n) \to 0$, there is N such that if $n \geq N$, then $(b_n - a_n) < \delta$. We already know that $a_n \leq A \leq b_n$. Using that $b_n - a_n < \delta$, we compute
$$b_n < a_n + \delta \leq A + \delta$$
Also, $(b_n - a_n) < \delta$ gives $b_n - \delta < a_n$, and so
$$a_n > b_n - \delta \geq A - \delta$$
We see that $A - \delta < a_n < b_n < A + \delta$, and (c) holds. □

4. Problems

1. Find an interesting pattern in the sequence defined by $a_0 = -3$, $a_1 = -6$, and $a_{n+2} = 4 \cdot a_{n+1} - 4 \cdot a_n$ for $n = 0, 1, 2, \ldots$. how many are?

2. Define
$$b_1 = 1 \quad \text{and} \quad b_2 = 1 \quad \text{and} \quad b_{n+2} = \frac{1}{n} \cdot b_{n+1} + b_n \quad \text{for} \quad n \geq 1$$
Show that $b_{2k-1} = b_{2k}$ for all integers $k \geq 1$. (Hint: induction.)

3. Show that the informal definition of $\binom{\alpha}{4}$ agrees with the formal definition. (Use the recursion in the formal definition to get an expression for $\binom{\alpha}{4}$.)

4. Find a nice formula for $\binom{-1}{n}$.

5. Prove the following famous and crazy formula:
$$\binom{-1/2}{n} = (-1)^n \cdot \frac{(2n)!}{4^n \cdot (n!)^2} \quad \text{for} \quad n = 0, 1, 2, \ldots$$
induction

6. Show that the sequence $(5n+1)/(17n-2)$ is eventually less than $3/10$.

7. Suppose that $|a_k| \to 0$ as $k \to \infty$. Prove that $a_k \to 0$ as $k \to \infty$.

8. Let k be a positive integer. Show carefully that $\frac{n^k}{2^n} \to 0$ as $n \to \infty$. (Hint: Monotone Convergence.)

9. Define
$$a_n = \sum_{k=1}^{n} \frac{1}{\sqrt[3]{8 \cdot n^3 + k}}$$
Prove that $a_n \to 1/2$ as $n \to \infty$. (Hint: estimate the sequence above and below. Note that n is constant in the sum, but k varies there.)

10. Show that the sequence $n^2 - 13n - 5$ is eventually strictly increasing.

11. Show that the sequence $(7n+9)/(n^2 - 3)$ is eventually decreasing.

12. Let $\alpha > -1$. Show that the sequence $|\binom{\alpha}{n}|$ is eventually decreasing. (Note: the sequence terms can become constant!)

13. Assuming the existence of $\sqrt{\alpha}$ when $\alpha > 1$, show that $\sqrt{\alpha}$ exists when $0 \leq \alpha \leq 1$.

14. Let x be a positive constant, and define $a_n = x^n/n!$. Show that a_n is eventually decreasing. Argue that it has a limit and find that limit. induction?

15. Given the number $\alpha > 1$, define the sequence a_n where $a_0 = 1$ and $a_{n+1} = (\alpha \cdot a_n)^{1/4}$ for $n \geq 0$. Prove: if $a_n^3 < \alpha$ (as is true when $n = 0$), then $a_n < a_{n+1}$ and $a_{n+1}^3 < \alpha$. Conclude that the sequence is increasing and bounded. Show that the cube of the limit is α. (Thus, we have constructed $\alpha^{1/3}$.)

16. Define $a_0 = 0$ and
$$a_{n+1} = \frac{1}{3 + a_n} \quad \text{for} \quad n \geq 0$$
Experiment with a_n to conjecture:

a) Is a_n increasing? decreasing? neither?

b) Does a_n seem to have a limit?

c) Assuming that a_n has a non-zero limit, find that limit.

17. Suppose that a_n is defined for $n \geq 1$ and that $a_n \to 0$ as $n \to \infty$. Define
$$b_n = \frac{1}{n} \cdot \sum_{j=1}^{n} a_j \quad \text{for} \quad n \geq 1$$

Complete the following idea to prove that $b_n \to 0$ as $n \to \infty$: Let $\epsilon > 0$; get N such that if $n \geq N$ then $|a_n| < \epsilon$. For $n \geq N$, estimate b_n by dividing the sum into two parts: $1 \leq j \leq N$ and $j > N$. Let $n \to \infty$ in each of the parts to get an upper bound on $|b_n|$.

18. Define $F_0 = 2$ and $F_{n+1} = 2 + 1/F_n$ for $n = 0, 1, 2 \ldots$. Show that F_n bounces back and forth on either side of its limit. (Hint: assuming there's a limit, find it first. Eventually you will need to prove your guess about the limit.)

CHAPTER 3

Limit Points and Function Limits.

1. Limit Points and BW.

Given a subset of the real numbers, a *limit point* of the set is a number that has elements of the subset arbitrarily close to it. Here is the formal definition: Let $S \subseteq \mathbb{R}$ and let $p \in \mathbb{R}$; then p is a *limit point* of S, if for all open intervals I with $p \in I$, the set $I \cap S$ has at least two elements. The term *accumulation point* is sometimes used for *limit point*. Notice that a limit point is **not** required to be an element of the set in question.

The definition requires there to be at least two elements in every open interval; actually there have to be infinitely many.

PROPOSITION 3.1. *Let $S \subseteq \mathbb{R}$ and $p \in \mathbb{R}$. Then p is a limit point of S if and only if every open interval containing p contains infinitely many elements of S.*

PROOF. First, assume there are $a, b \in \mathbb{R}$ with $a < p < b$, and $(a, b) \cap S$ is finite. Call this set F, and then the set $F \cup \{a, b, p\}$ is also finite, and so we can put its elements in order:

$$a = f_1 < f_2 < \cdots < f_n = b$$

Now $p = f_k$ for some k, and since $a < p < b$, we see that $k > 1$ and $k < n$. Then $(f_{k-1}, f_{k+1}) \cap S$ is contained in the set $\{f_k\}$, and so this set has 0 or 1 elements. This shows that p is not a limit point of S.

The contrapositive of what we just proved: if p is a limit point of S and $a < p < b$, then the set $(a,b) \cap S$ is infinite.

Conversely, if every open interval containing p contains infinitely many elements of S, then every open interval containing p certainly contains at least two elements of S! Thus, p is a limit point of S. □

Proposition 3.1 shows that a finite set cannot have limit points. We will see that the integers form an *infinite* set with no limit points.

If a set is contained in a closed interval, its limit points are contained there as well.

*PROPOSITION 3.2. *If $S \subseteq \mathbb{R}$ is contained in the closed interval $[a,b]$, and if p is a limit point of S, then $p \in [a,b]$.*

PROOF. We will show that if $p \notin [a,b]$, then p is not a limit point of S. For instance, if $p < a$, then the open interval $(p-1, a)$ contains p but no element of S. The proof for the case $b < p$ is similar. □

We come to our main theorem about limit points: the *Bolzano-Weierstrass Theorem*, known affectionately as BW. This theorem turns out to be equivalent to the Completeness Property of the real numbers.

For many theorems in the text, we will have two general proof techniques. One technique will involve bisecting intervals, over and over, to narrow down on a number of interest. The second technique will employ Completeness directly – usually in a clever way. The next two proofs will illustrate the two techniques. For each proof, there will be a way to use the other technique, as well. We will use the text, class, and homework to give examples of the usage as we go through the next couple of chapters. We hope you understand that we are not just interested in facts such as BW but in general proof techniques typical of analysis. Thus, it may well be a good idea to have more than one proof of a given fact, since different proofs highlight different analytic ideas.

BOLZANO-WEIERSTRASS THEOREM. *Every bounded infinite subset of the reals has at least one limit point.*

PROOF. Let S be a bounded, infinite subset of the reals. Then S is contained in some closed interval $S \subseteq [a,b]$, with $a < b$.

We will use repeated bisection and Proposition 2.7. Define $a_0 = a$ and $b_0 = b$. There are infinitely many elements of S in $[a_0, b_0]$, since S is infinite. Let $c = (a_0 + b_0)/2$. Since $[a_0, b_0]$ is union of $[a_0, c]$ and $[c, b_0]$, one of these two sub-intervals must contain infinitely many elements of S. (Both could as well.) If $[a_0, c]$ contains infinitely many elements of S, then define $a_1 = a_0$ and $b_1 = c$. Otherwise, define $a_1 = c$ and $b_1 = b_0$. In any case, $[a_1, b_1]$, whose width is $(b_0 - a_0)/2$, contains infinitely many elements of S.

Keep going. Assume we have constructed
$$a_0 \le a_1 \le \cdots a_n < b_n \le b_{n-1} \le \cdots \le b_0$$
where $[a_n, b_n]$ has width $(b_0 - a_0)/2^n$ and contains infinitely many elements of S. Let $c = (b_n + a_n)/2$, and if $[a_n, c]$ contains infinitely many elements of S, then define $a_{n+1} = a_n$ and $b_{n+1} = c$. Otherwise, define $a_{n+1} = c$ and $b_{n+1} = b_n$. In any case, $[a_{n+1}, b_{n+1}]$, whose width is $(b_0 - a_0)/2^{n+1}$, contains infinitely many elements of S.

Because of the ordering of the sequences, and since
$$b_n - a_n = (b_0 - a_0)/2^n \to 0 \quad \text{as} \quad n \to \infty$$
Proposition 2.7 holds, and it finds a limit A of the sequence a_n.

We claim that A is a limit point of S. To see this, suppose that $c, d \in \mathbb{R}$ with $c < A < d$. Proposition 2.7c finds n such that $[a_n, b_n] \subset (c, d)$. The interval $[a_n, b_n]$ contains infinitely many elements of S, and so $(c, d) \cap S$ is infinite as well. □

We mentioned previously that a finite set has no limit points, so it is necessary to BW that the set is infinite.

2. Consequences of BW.

We go back to Chapter 2 to discuss the idea of knowing that a sequence converges *without knowing the limit in advance*. Cauchy found a way to do this; in his honor we will say that the sequence a_n is a *Cauchy sequence* if for every $\epsilon > 0$, there is an integer N such that if $n, m > N$, then $|a_n - a_m| < \epsilon$. The point of the following proposition is that the Cauchy sequences are exactly the converging sequences. We will not make much use of Proposition 3.3 in the course; we may have time to use it in examples worked out in class. But the idea of a Cauchy sequence is important in that it generalizes to many other contexts, and it will be advantageous for you to give it a first look here.

PROPOSITION 3.3. *Let a_n be a sequence of real numbers, with $n \geq 0$. Then a_n converges if and only if it is a Cauchy sequence.*

PROOF. Suppose that $a_n \to A$ as $n \to \infty$. Let $\epsilon > 0$. There is N such that if $n \geq N$, then $|a_n - A| < \epsilon$. For $n, m \geq N$, we estimate

$$|a_n - a_m| = |a_n - A + A - a_m| \leq |a_n - A| + |A - a_m| < 2 \cdot \epsilon$$

This proves that a_n is a Cauchy sequence.

Now suppose that a_n is a Cauchy sequence. We claim first that a_n is bounded. Taking $\epsilon = 1$, get N such that if $n, m > N$, then $|a_n - a_m| < 1$. Choose a particular $m > N$, and then for all $n > N$, we have

$$|a_n| = |a_n - a_m + a_m| \leq |a_n - a_m| + |a_m| < 1 + |a_m|$$

Thus, the a_n with $n > N$ are bounded, and it follows that the entire sequence is bounded.

Next, we claim that there is a real number p such that for all $\epsilon > 0$, there are infinitely many k such that $|p - a_k| < \epsilon$. Indeed, assume first that there are infinitely many values a_k. (Not infinitely many k, but infinitely many a_k.) Then BW gives the set of values a_k a limit point p. If $\epsilon > 0$, then

Proposition 3.1 shows that the open interval $(p-\epsilon, p+\epsilon)$ contains infinitely many values of a_k. These infinitely many values must involve infinitely many subscripts, and so there are infinitely many k such that $|p-a_k| < \epsilon$, as claimed.

To complete the proof of the claim, we need to assume that there are only finitely many values a_k. Then there is at least one such value p that comes from infinitely many subscripts. We mean that $p = a_k$ for infinitely many k. The claim holds again.

Given the number p from the claim, we now show that $a_k \to p$ as $k \to \infty$, so that a_k converges. This will complete the proof. Let $\epsilon > 0$. Since a_k is a Cauchy sequence, there is N such that if $n, m > N$, then $|a_n - a_m| < \epsilon$. There are infinitely many k such that $|p - a_k| < \epsilon$, and so there is such a $k > N$. Now we show that $m > N$ implies that $|p - a_m| < 2\epsilon$. Indeed, for such an m, we compute

$$|p - a_m| = |p - a_k + a_k - a_m| \le |p - a_k| + |a_k - a_m| < \epsilon + \epsilon$$

as needed. □

We turn to another theorem which generalizes extensively. The *Heine-Borel Theorem* is related to BW in being a fundamental use of completeness; in fact, they are often seen as interchangeable. We will probably not have time to discuss this theorem in class, but it is a commonly used tool in advanced work. The theorem involves open intervals around each element of a closed interval, and it asserts that the closed interval can be covered by finitely many of the open intervals. In technical terms, the theorem says that each closed interval is *compact*.

3. LIMIT POINTS AND FUNCTION LIMITS.

THE HEINE-BOREL THEOREM ON THE REALS. *Let $a < b$ be real numbers, and suppose that for each $c \in [a,b]$, there is an open interval I_c with $c \in I_c$. Then there is a finite set $F \subset [a,b]$ such that*

$$[a,b] \subseteq \bigcup_{x \in F} I_x$$

PROOF. Suppose that J is a subset of $[a,b]$. A *finite cover* for J is a finite subset G of $[a,b]$, such that

$$J \subseteq \bigcup_{x \in G} I_x$$

To repeat: a finite cover is a finite subset of $[a,b]$ whose open intervals cover (contain) J. For instance, if $J = \{c\}$, with $c \in [a,b]$, then $\{c\}$ is a finite cover of J, since $c \in I_c$. Similarly, every finite subset of $[a,b]$ has a finite cover. The proof involves a sophisticated way of producing finite covers for larger sets.

Let S be the set of $x \in [a,b]$ such that $[a,x]$ has a finite cover. We claim that $a \in S$. Indeed, $a \in I_a$, and we see that $\{a\}$ is a finite cover for $[a,a] = \{a\}$. In particular, S is not empty. The definition of S shows that it is bounded above by b, and so Completeness gives S a sup s. Since I_s is an open interval, Proposition 1.1b finds $t \in S$ with $t \in I_s$ and $t \leq s$.

We claim that $s \in S$. Indeed, we have $t \in S$, and so $[a,t]$ has a finite cover: say that G is a finite subset of F with

$$[a,t] \subseteq \bigcup_{x \in G} I_x$$

Since $t \in I_s$, we see that

$$[a,s] = [a,t] \cup [t,s] \subset \left[\bigcup_{x \in G} I_x\right] \cup I_s$$

The set $G \cup \{s\}$ is finite, and we see that $[a,s]$ has a finite cover. Thus, $s \in S$.

We claim that $s = b$. If $s < b$, then since I_s is an open interval, there is $c \in I_s$ with $s < c < b$. There is a finite cover of $[a,s]$, and we can include s in the cover, so that I_s is one of the covering sets. Since $[s,c] \subset I_s$, we see that the interval $[a, c]$ is contained in the cover union. Thus, $c \in S$. But $s < c$ and s is an upper bound for S. This is a contradiction. We conclude that $s = b$.

We have proved that $b \in S$. In other words, $[a, b]$ has a finite cover, and that is what our theorem asserts. □

3. Function Limits.

You may have the impression that calculus is the mathematics of function limits. While it is true that function limits unify much of the calculus, we prefer to say that calculus is the mathematics of continuity. We will see that the function limit one encounters in calculus is deficient and we will avoid it whenever we can. For instance, both the derivative and the integral can be defined without the traditional function limit. So why, you may ask, are we introducing function limits at all? There are at least two main reasons: 1) an important goal of this course is to make formal sense of the calculus you have seen previously; 2) the proofs of limit facts will involve limit points, and we want to reinforce the properties of limit points.

The notation $x \to \alpha$ familiar from limits will arise when x is an element of the domain of a function, and we will see that we need α to be a limit point of that set. This is why we began this chapter with limit points rather than with limits.

So, here is the setting. We have a subset S of the reals, and a function $f : S \to \mathbb{R}$. Suppose that α is a limit point of S, and that $A \in \mathbb{R}$. We say that $f(x) \to A$ as $x \to \alpha$ if for every $\epsilon > 0$, there is a positive number δ such that if $x \in S$ and $0 < |x - \alpha| < \delta$, then $|f(x) - A| < \epsilon$. Notice that in order for f to have a limit on S, it is necessary for S to have a limit point.

The foregoing will need close study. Here is a less formal way to express it: we can make $f(x)$ arbitrarily close to A by making x close to α. The number ϵ measures how close $f(x)$ needs to be to A. To say that ϵ is arbitrary is to say that $f(x)$ can be made *arbitrarily close* to A. The number δ measures how close x needs to be to α to make $f(x)$ close to A.

You might recall the condition $0 < |x - \alpha|$ in the inequality $0 < |x - \alpha| < \delta$. This says that $x \ne \alpha$; we need this condition in case $f(\alpha)$ is not defined or is defined but not equal to A. The limit is about where the function is *going* as $x \to \alpha$, not about where it *arrives* when $x = \alpha$. This last condition is a headache, as we will see.

Let's do an example – although we warn you that the definition of the limit is of more *logical* use than *computational* use. We will get to the familiar limit algebra very soon. But let's try to use the definition to prove that

$$\frac{x^2 + 2}{x - 1} \to \frac{11}{2} \quad \text{as} \quad x \to 3$$

(Of course, we obtained the limit by "plugging in." We'll explain this in the next chapter.) Notice that the domain of the function on the left is $\mathbb{R} \setminus \{1\}$, and 3 is a limit point of this set. We need to find $\delta > 0$ so that the quantity

$$\left| \frac{x^2 + 2}{x - 1} - \frac{11}{2} \right|$$

is small. Intuitively, this should be true because $x = 3$ was used to get the right side. Let's try to estimate this difference.

$$\left| \frac{x^2 + 2}{x - 1} - \frac{11}{2} \right| = \left| \frac{2(x^2 + 2) - 11(x - 1)}{2(x - 1)} \right| = \left| \frac{2x^2 - 11x + 15}{2(x - 1)} \right|$$
$$= \left| \frac{(2x - 5)(x - 3)}{2(x - 1)} \right| = |x - 3| \cdot \left| \frac{2x - 5}{2x - 2} \right|$$

We will have $|x - 3| < \delta$, and this ought to make the last expression small. To be sure that this is the case, we bound the second factor on the right. There are many ways to do this – the particulars are less important than obtaining a bound. For instance, we want $x \approx 3$ and we need to stay away from $x = 1$. So we might make sure that $\delta \le 1$, so that $|x - 3| < \delta$ gives $2 < x < 4$. (Actually, we will probably have δ quite a bit smaller, the present condition $\delta \le 1$ is

meant to help bound the fraction above.) With $2 < x < 4$, we get the crude bound
$$\left|\frac{2x-5}{2x-2}\right| \leq \frac{2\cdot 4 - 5}{2\cdot 2 - 2} = \frac{3}{2}$$
and then we have established this implication:

$$\text{If } |x-3| < \delta \leq 1 \text{ then } \left|\frac{x^2+2}{x-1} - \frac{11}{2}\right| \leq \delta \cdot \frac{3}{2}$$

The far right side is less than an arbitrary ϵ if $\delta \leq 2\epsilon/3$. We have established the desired limit. We end with a disclaimer: even though the foregoing calculation is very much in the spirit of typical analysis, we will not make a living calculating limits this way.

As we did with sequences, we establish that the limit is unique if it exists, and we prove that the limit obeys the expected algebraic properties. The following proofs exhibit a more typical use of the limit definition than we saw in the particular example.

*PROPOSITION 3.4. *Suppose that $S \subseteq \mathbb{R}$ and $f : S \to \mathbb{R}$ and α is a limit point of S and $A, B \in \mathbb{R}$ and $f(x) \to A$ and $f(x) \to B$ as $x \to \alpha$. Then $A = B$.*

PROOF. Suppose that $f(x) \to A$ and $f(x) \to B$. Let $\epsilon > 0$. Because $f(x) \to A$, there is a positive number δ_1 such that if $x \in S$ and $0 < |x-\alpha| < \delta_1$, then $|f(x) - A| < \epsilon$. Because $f(x) \to B$, there is a positive number δ_2 such that if $x \in S$ and $0 < |x - \alpha| < \delta_2$, then $|f(x) - B| < \epsilon$. Let δ be the smaller of δ_1, δ_2, and we have this: if $x \in S$ and $0 < |x - \alpha| < \delta$, then $|f(x) - A| < \epsilon$ and $|f(x) - B| < \epsilon$.

Since α is a limit point of S, the open interval $(\alpha - \delta, \alpha + \delta)$ contains at least two elements of S. These two elements cannot both be α (maybe neither of them is α); so we can choose $x \in S$ with $\alpha - \delta < x < \alpha + \delta$ and with $x \neq \alpha$. Then $0 < |x - \alpha| < \delta$, and we conclude that $|f(x) - A| < \epsilon$ and $|f(x) - B| < \epsilon$.

Compute
$$|A - B| = |A - f(x) + f(x) - B| \leq |A - f(x)| + |f(x) - B| < 2\epsilon$$
Because ϵ is arbitrary, we see that $A = B$. $\square$

Here is an example that shows why we need α to be a limit point of the domain S. Let $f : [0,1] \cup \{2\} \to \mathbb{R}$ be defined by $f(x) = x$. If we allow $x \to 2$, let's show that $f(x) \to A$ where A is an *arbitrary number*. Given $\epsilon > 0$, take $\delta = 1$, and then the hypothesis
$$x \in \left([0,1] \cup \{2\}\right) \quad \text{and} \quad 0 < |x - 2| < 1$$
is not satisfied by any x. In other words, the hypothesis is *vacuous*; an implication with a vacuous hypothesis is true by definition, and so the implication
$$\textbf{If } \left[x \in \left([0,1] \cup \{2\}\right) \text{ and } 0 < |x - 2| < 1\right] \textbf{then } |f(x) - A| < \epsilon$$
is true. This specific example symbolizes the general case: if α is not a limit point of the domain of the function f, then the limit of $f(x)$ as $x \to \alpha$ would be arbitrary if it were allowed at all.

In the case of a unique limit, we can refer to the limit by the customary limit that just mentions $f(x)$ and α.
$$\lim_{x \to \alpha} f(x) = A \quad \text{means} \quad f(x) \to A \text{ as } x \to \alpha$$
Proposition 3.4 validates this notation, which expresses the limit in terms of the function and α.

Observe that the domain of f has an effect on the limiting value. For instance, define $f(x) = 0$ when $x < 0$ and $f(x) = 1$ for $x \geq 0$. As a function on $[0,1]$, we have $f(x) \to 1$ as $x \to 0$. As a function on $[-1, 0]$, we have $f(x) \to 0$ as $x \to 0$. (Notice that $f(0) = 1$ is irrelevant to the second limit, since $x \to 0$ never "looks at" $x = 0$.) As a function on $[-1, 1]$, $f(x)$ has no limit as $x \to 0$. In almost all limit calculations, the domain is understood, but you should check each time.

3. FUNCTION LIMITS.

One-sided limits are automatically included in our definition, for if $f(x)$ is defined, say, for $x \in (a, b]$, then $\lim_{x \to a} f(x)$ is automatically $x \to a+$, because the definition of limit requires x to be chosen from the domain $(a, b]$. Similarly, if $f(x)$ is defined on $[c, a)$, then $x \to a$ means $x \to a-$.

You know the limit algebra. With a more general domain, you just have to be sure that all the functions involved in a limit calculation have a large enough common domain. We'll see how much of the proof of the following we have time for in class or on homework. We have included part of the proof of statement (c).

PROPOSITION 3.5. *Suppose that $S \subseteq \mathbb{R}$, and that $f, g : S \to \mathbb{R}$. Let α be a limit point of S, and suppose that $f \to A$ and $g \to B$ as $x \to \alpha$. Then we have the following.*

a) $(f(x) + g(x)) \to (A + B)$ *as* $x \to \alpha$.
b) $(f(x) \cdot g(x)) \to (A \cdot B)$ *as* $x \to \alpha$.
**c)* *If $B \neq 0$, then the domain of $f(x)/g(x)$ has α as a limit point, and $(f(x)/g(x)) \to (A/B)$ as $x \to \alpha$.*

PROOF. For (c), we will prove that $1/g(x) \to (1/B)$ as $x \to \alpha$. Let T be the set of $x \in S$ such that $g(x) \neq 0$. We will show that α is a limit point of T; it takes a couple of steps.

Using $|B|/2$ as an "epsilon" we find $\delta > 0$ such that if $0 < |x - \alpha| < \delta$ and $x \in S$, then $|g(x) - B| < |B|/2$. For such x, use the reverse triangle inequality to show that

$$|g(x)| = |B - (B - g(x))| \geq |B| - |B - g(x)| > |B| - \frac{|B|}{2} = \frac{|B|}{2}$$

In particular, for these x, we have $g(x) \neq 0$, and so $x \in T$. Let

$$J = (\alpha - \delta, \alpha + \delta) \setminus \{\alpha\}$$

and we have shown that $S \cap J \subseteq T$.

Now let I be an open interval with $\alpha \in I$ and we will show that $I \cap T$ is infinite. We know that $I \cap (\alpha - \delta, \alpha + \delta)$ is an open interval containing α, and so it has infinite intersection with S, since α is a limit point of S. It follows that
$$I \cap J \cap S = I \cap (\alpha - \delta, \alpha + \delta) \cap S \setminus \{\alpha\}$$
has infinitely many elements. But $I \cap J \cap S \subseteq I \cap T$, and so $I \cap T$ is infinite, as needed.

As to the limit of $1/g(x)$, let $\epsilon > 0$ and get $\mu > 0$ such that if $x \in S$ and $0 < |x - \alpha| < \mu$, then $|g(x) - B| < \epsilon$. We can assume that $\mu \leq \delta$, and then $|g(x)| > |B|/2$, as shown above. Then
$$\left| \frac{1}{g(x)} - \frac{1}{B} \right| = \left| \frac{B - g(x)}{g(x) \cdot B} \right| = \frac{|B - g(x)|}{|g(x)| \cdot |B|} < \frac{\epsilon}{|B| \cdot |B|/2}$$
This quantity can be made arbitrarily small. $\square$

Recall the analogous result for sequences: Proposition 2.4. We pointed out that this result includes the case of constant multiples of sequences. So it is also the case here: if $g(x) = \beta$, a constant function, then Proposition 3.5b shows that if $f(x) \to A$, then $\beta \cdot f(x) \to \beta \cdot A$.

Above, we did a limit from the definition:
$$\lim_{x \to 3} \frac{x^2 + 2}{x - 1} = \frac{11}{2}$$
Proposition 3.5 allows us to do this limit algebraically (as we are used to doing):
$$\lim_{x \to 3} \frac{x^2 + 2}{x - 1} = \left[\left(\lim_{x \to 3} x \right) \cdot \left(\lim_{x \to 3} x \right) + \lim_{x \to 3} 2 \right] \cdot \frac{1}{\lim_{x \to 3} x - \lim_{x \to 3} 1}$$
$$= [3 \cdot 3 + 2] \cdot \frac{1}{3 - 1} = \frac{11}{2}$$
It is an important detail in such standard calculations to notice that it is the existence of the limits on the right side that implies the existence of the limit

on the left. In other words, Proposition 3.5 *does not say*, for instance, that

$$\text{(3.1)} \qquad \lim_{x \to \alpha} \left[f(x) \cdot g(x) \right] = \lim_{x \to \alpha} f(x) \cdot \lim_{x \to \alpha} g(x)$$

For notice that $1 = x \cdot 1/x$ for all $x \neq 0$, but

$$\lim_{x \to 0} 1 \neq \lim_{x \to 0} x \cdot \lim_{x \to 0} \frac{1}{x}$$

since the second limit on the right does not exist. What Proposition 3.5 does say is that (3.1) is true *if the limits on the right side exist.*

We turn to polynomials. The following can be proved using Proposition 3.5, since polynomials are built up using ordinary arithmetic. Our proof, however, uses factoring.

*PROPOSITION 3.6. *Let $f(x)$ be a polynomial, and let $\alpha \in \mathbb{R}$. Then $f(x) \to f(\alpha)$ as $x \to \alpha$.*

PROOF. Notice that the domain of $f(x)$ is $\mathbb{R}$, and so α is necessarily a limit point of this domain. The division theorem of algebra finds a polynomial $g(x)$ such that $f(x) = f(\alpha) + (x - \alpha) \cdot g(x)$. Choose $\delta > 0$, and since $g(x)$ is a polynomial, then an exercise we did finds a positive number M such that $|g(x)| \leq M$ for all $x \in [\alpha - \delta, \alpha + \delta]$. For these x, we see that

$$|f(x) - f(\alpha)| \leq M \cdot |x - \alpha| \leq M \cdot \delta$$

Given $\epsilon > 0$, let $\delta = \epsilon/M$, and will have the limit implication. $\square$

Limits do not behave well under function composition, as we will now see. In the following example, it might help to graph the functions. Define $f(x) = 0$ for $x \neq 0$ and define $f(0) = 1$. It is easy to see that $f(x) \to 0$ as $x \to 0$. Now suppose we intend to have $x = g(t)$ where $g(t) = 0$ for all t. Then certainly $g(t) \to 0$ as $t \to 0$. In other words $x \to 0$ as $t \to 0$. It would seem that we could chain the limits: as $t \to 0$, we have $x \to 0$, and as $x \to 0$, we have

$f(x) \to 0$. But this is false, since $f(x) = f(g(t)) = f(0) = 1$, and so as $t \to 0$, we have $f(x) \to 1$.

More complicated examples are possible; the upshot is that function limits do not in general persist over function composition. Function composition comes up a great deal in analysis and so the failure of limits in composition is a severe deficiency. That is why we said, at the beginning of this section, that our work with derivatives and integrals will avoid the function limit and work more cleverly.

4. Infinity.

We want to define what is meant by $x \to \infty$ and/or $f(x) \to \infty$. Along the way, we hope to justify familiar calculations, such as

$$\lim_{x \to 0+} \frac{1}{x} = \infty \quad \text{and} \quad \lim_{x \to \infty} \frac{1}{x} = \frac{1}{\infty} = 0$$

We realize that infinity is not a number and cannot simply be included in arbitrary calculations. For instance, you have seen that ∞/∞ in a limit calculation can be absolutely anything, including undefined.[1]

Some people feel that a first course in analysis should avoid infinity to keep the focus on real number calculations. Frankly, that is what we will do for the most part. But infinity has already appeared in the formulation of sequence limits: $a_k \to A$ as $k \to \infty$, and so the cat is out of the bag, so to speak. We do want to give a formal introduction to limits where the variable and/or the function goes to infinity, with the understanding that this is a sidelight to our course.

Consider the limit implication: "If $0 < |x - \alpha| < \delta$, then $|f(x) - A| < \epsilon$." This implication will not make sense in the case that $x \to \infty$ or $x \to -\infty$ or in the case that $f(x) \to \infty$ or $f(x) \to -\infty$, since we cannot put infinity into an inequality, in place of α and/or A, as if it were a number. The way out of

---

[1]Consider these ratios as $x \to \infty$: $3x/x$ and x^2/x and x/x^3 and $x(2 + \sin(x))/x$.

this problem is to define the limit in such a way as not to mention $x - \alpha$ or $f(x) - A$ directly. The inequality $|x - \alpha| < \delta$ is meant to say "x is close to α." The number δ measures the closeness to α, and we have in mind that δ is small. Analogously, to say that "x is close to ∞" is to say that "x is large." This is expressed formally by saying that $x > \delta$ for a number δ. We have in mind that the number δ is large, but as in the case of $|x - \alpha| < \delta$, "small" and "large" are relative. We need to understand the properties of the limit that follow from the inequality in the definition.

Similarly, to say "x is close to $-\infty$" is to say $x < \delta$ for some δ. (And we have in mind that δ is a "large negative number.")

The notation $\lim_{x \to \alpha} f(x) = A$ implies that α is a limit point of the domain of f. If $f : S \to \mathbb{R}$, then ∞ is a limit point of S if S is **not** bounded above. We have that $-\infty$ is a limit point of S if S is **not** bounded below.

With all this in mind, here are the eight (!) relevant limit definitions. In each case $S \subseteq \mathbb{R}$ and $f : S \to \mathbb{R}$.

(1) $\lim_{x \to \infty} f(x) = A$ means: S is not bounded above and for all $\epsilon > 0$ there is δ such that if $x \in S$ and $x > \delta$, then $|f(x) - A| < \epsilon$.

(2) $\lim_{x \to \infty} f(x) = \infty$ means: S is not bounded above and for all ϵ there is δ such that if $x \in S$ and $x > \delta$, then $f(x) > \epsilon$.

(3) $\lim_{x \to \infty} f(x) = -\infty$ means: S is not bounded above and for all ϵ there is δ such that if $x \in S$ and $x > \delta$, then $f(x) < \epsilon$.

(4) $\lim_{x \to -\infty} f(x) = A$ means: S is not bounded below and for all $\epsilon > 0$ there is δ such that if $x \in S$ and $x < \delta$, then $|f(x) - A| < \epsilon$.

(5) $\lim_{x \to -\infty} f(x) = \infty$ means: S is not bounded below and for all ϵ there is δ such that if $x \in S$ and $x < \delta$, then $f(x) > \epsilon$.

(6) $\lim_{x \to -\infty} f(x) = -\infty$ means: S is not bounded below and for all ϵ there is δ such that if $x \in S$ and $x < \delta$, then $f(x) < \epsilon$.

(7) $\lim_{x \to \alpha} f(x) = \infty$ means: α is a limit point of S and for all ϵ there is $\delta > 0$ such that if $x \in S$ and $0 < |x - \alpha| < \delta$, then $f(x) > \epsilon$.

(8) $\lim_{x \to \alpha} f(x) = -\infty$ means: α is a limit point of S and for all ϵ there is $\delta > 0$ such that if $x \in S$ and $0 < |x - \alpha| < \delta$, then $f(x) < \epsilon$.

In exercises below you are asked to established some typical formulas.

5. Problems

1. What are the limit points of the half-open interval $[0, 1)$?

2. Prove that the integers have no limit points.

3. Every real number is a limit point of the rational numbers. (Hint: density!)

4. Let S, T be subsets of the reals with $S \subseteq T$. If p is a limit point of S, then p is a limit point of T.

5. Let S, T be subsets of the reals. If p is a limit point of $S \cup T$, then p is a limit point of S or p is a limit point of T. (Hint: assume that p is not a limit point of T – what does that mean?)

6. Give an example of subsets S, T of the reals, and $\alpha \in \mathbb{R}$ such that α is a limit point of S, and α is a limit point of T, and α is **not** a limit point of $S \cap T$.

7. Let S be a subset of the reals, and let L be the set of limit points of S. If p is a limit point of $S \cup L$, then p is a limit point of S. (Hint: use the literal definition of *limit point* carefully.)

8. Complete the following alternative proof of BW. Given the infinite set $S \subseteq [a, b]$, define T to be the set of $x \in [a, b]$ such that $[a, x] \cap S$ is finite. Show that T is non-empty and bounded above, and let c be the sup of T. Show that either $c < b$ or $c \notin T$. In either case, show that c is a limit point of S.

5. PROBLEMS

9. Complete the following alternative proof of the Heine-Borel Theorem. Assume that we have open intervals I_c for all $c \in [a, b]$. We work by contradiction, assuming $[a, b]$ has no finite cover. Define $a_0 = a$ and $b_0 = b$. Let c be the midpoint of $[a, b]$. Show that if $[a, c]$ and $[c, b]$ have finite covers, then so does $[a, b]$, a contradiction. If $[a, c]$ does not have a finite cover, define $a_1 = a$ and $b_1 = c$. Otherwise, define $c = a_1$ and $b = b_1$. Keep going, bisecting your way to a contradiction!

10. Use the definition of function limit to prove that $x^2 \to 9$ as $x \to 3$.

11. Let $L > 0$ and $f : \mathbb{R} \to \mathbb{R}$ with $f(x) \to L$ as $x \to 1$. Use the definition of function limit to show that $\sqrt{f(x)} \to \sqrt{L}$ as $x \to 1$.

12. Show that $1/x \to \infty$ as $x \to 0+$.

13. Let $f(x)$ be a non-constant polynomial. Show that either $f(x) \to \infty$ or $f(x) \to -\infty$ as $x \to \infty$, and identify the particular polynomials that have the particular limits. (Hint: prove first that $1/x^k \to 0$ as $x \to \infty$, where k is a positive integer.)

14. Let $S \subseteq \mathbb{R}$ and $f : S \to \mathbb{R}$, and suppose that $f(x) \to \infty$ as $x \to \alpha$. Show that $1/f(x) \to 0$ as $x \to \alpha$. (Note: one thing you need to do is to show that α is a limit point of the set of $x \in S$ such that $f(x) \neq 0$.)

15. Let $S \subseteq \mathbb{R}$ not be bounded above. Let $f : S \to \mathbb{R}$ and $g : S \to \mathbb{R}$ and let $f(x) \to 7$ as $x \to \infty$, and let $g(x) \to -2$ as $x \to \infty$. Show that $f(x) + g(x) \to 5$ as $x \to \infty$.

16. Let $S \subseteq \mathbb{R}$, let α be a limit point of S, let $f : S \to \mathbb{R}$ and $g : S \to \mathbb{R}$, and suppose that $f(x) \to \infty$ as $x \to \alpha$ and that $g(x) \to -2$ as $x \to \alpha$. Show that $f(x) \cdot g(x) \to -\infty$ as $x \to \alpha$. (Hint: First get x close enough to α so that $g(x) < -1$.)

17. Let $f(x)$ and $g(x)$ be polynomials of the same degree. Show that
$$\frac{f(x)}{g(x)} \to \frac{a}{b} \quad \text{as} \quad x \to \infty$$
where a is the leading coefficient of f and b the leading coefficient of g.

CHAPTER 4

Continuous Functions.

1. The Definition.

You know the definition: $f(x)$ is continuous at α if

(4.1) $$\lim_{x \to \alpha} f(x) = f(\alpha)$$

Since we have defined the general function limit in the previous section, we are in a position to make sense of the limit just given. But the deficiencies of the function limit suggest that we scrap the limit definition in favor of something better.

Let S be a subset of the real numbers and let $f : S \to \mathbb{R}$. Let $\alpha \in S$. Then $f(x)$ is $\textit{continuous}$ at α if for all $\epsilon > 0$ there is $\delta > 0$ such that if $|x - \alpha| < \delta$ and $x \in S$, then $|f(x) - f(\alpha)| < \epsilon$.

This definition looks very much like the limit definition of the previous chapter, with the following nuances: we allow $x = \alpha$ and it is not necessary for α to be a limit point of S. The α $\textit{does}$ need to be a number, however, and since we reference $f(\alpha)$, the number α must be in the domain S of f.

Just to give the limit definition of continuity (4.1) one last look, we will prove the following in class.

PROPOSITION 4.1. $\textit{Let } S \textit{ be a subset of the real numbers and let } f : S \to \mathbb{R}.$ $\textit{Suppose that } \alpha \in S \textit{ and that } \alpha \textit{ is a limit point of } S. \textit{ Then } f(x) \textit{ is continuous at } \alpha \textit{ if and only if the limit (4.1) holds.}$

Here is the only use we will make of Proposition 4.1. Proposition 3.6 used the limit algebra to see that $f(x) \to f(\alpha)$ as $x \to \alpha$ for every α if $f(x)$ is a polynomial. Proposition 4.1 says that polynomials are continuous at every real number.

Proposition 4.1 says that the limit definition covers the case when α is a limit point of the domain. Let us see that when α is not a limit point, we get continuity for free.

PROPOSITION 4.2. *Let S be a subset of the real numbers and let $f : S \to \mathbb{R}$. Suppose that $\alpha \in S$ and that α is not a limit point of S. Then $f(x)$ is continuous at α.*

PROOF. There is an open interval I containing α that contains at most one element of S. Since $\alpha \in I \cap S$, it must be that α is the only element of S in I. There is a positive number δ such that the open interval $(\alpha - \delta, \alpha + \delta) \subseteq I$, and so α is the only element of S in $(\alpha - \delta, \alpha + \delta)$. Let $\epsilon > 0$ be given and use this δ. The hypothesis: $|x - \alpha| < \delta$ and $x \in S$ implies that $x = \alpha$, and so

$$f(x) - f(\alpha) = f(\alpha) - f(\alpha) = 0 < \epsilon$$

□

Similar to the limit algebra, we have continuity algebra. We will prove some of these facts in class and, perhaps, some on homework. For statement (c) we will use a proof that is very similar to the proof of Proposition 3.5 in Chapter 3.

PROPOSITION 4.3. *Let S be a subset of the real numbers and let $f : S \to \mathbb{R}$ and $g : S \to \mathbb{R}$. Suppose that $\alpha \in S$ and that f, g are continuous at α. Then*
a) $f(x) + g(x)$ *is continuous at α;*
b) $f(x) \cdot g(x)$ *is continuous at α;*
c) *if $g(\alpha) \neq 0$, then $f(x)/g(x)$ is continuous at α.*

Proposition 4.3 says that a ratio of polynomials is continuous wherever it is defined. Ratios of polynomials are called *rational functions*. Soon we will have the continuity of other standard functions; we will need a few more theorems, as well as knowledge of the derivative, before we can address this.

Recall that limits do not persist through function composition. Continuity does persist, and that fact is one reason for choosing the continuous functions as a fundamental object of study in analysis.

**PROPOSITION 4.4. *Let S, T be subsets of the reals, let $f : S \to T$ and $g : T \to \mathbb{R}$. If $\alpha \in S$ and f is continuous at α and g is continuous at $f(\alpha)$, then $g(f(x))$ is continuous at α.*

PROOF. Let $\epsilon > 0$. Since g is continuous at $f(\alpha)$, there is $\delta > 0$ such that if $y \in T$ and $|y - f(\alpha)| < \delta$, then $|g(y) - g(f(\alpha))| < \epsilon$. Since f is continuous at α, there is $\mu > 0$ such that if $x \in S$ and $|x - \alpha| < \mu$, then $|f(x) - f(\alpha)| < \delta$.

Let $x \in S$ and $|x - \alpha| < \mu$. Then $|f(x) - f(\alpha)| < \delta$, and $f(x) \in f(S) \subseteq T$. Thus, $|g(f(x)) - g(f(\alpha))| < \epsilon$. □

For many applications, we have functions that are continuous at every point in their domain. Such a function is called simply *continuous*. We have to be sure we know what domain is intended, however. The function

$$h(x) = \begin{cases} 1 & \text{if } x > 0 \\ 0 & \text{if } x \leq 0 \end{cases}$$

is not continuous on $\mathbb{R}$. It **is** continuous on the set P of positive reals. It is also continuous on $P \cup \{-1\}$. (Why?)

We want to present a rather wild example of a function, an example that will be used at several points in the term. We need a sequence r_k for $k \geq 1$ of numbers in $[0, 1]$; we need the sequence to be one to one, so that if $r_k = r_j$,

then $k = j$. Define
$$f(x) = \begin{cases} 1/k & \text{if } x = r_k \\ 0 & \text{otherwise} \end{cases}$$
We will call f the *Dirichlet function for the sequence r_k*. A very interesting case results when we let r_k be a sequence hitting all the rational numbers in $[0, 1]$. (The existence of such a sequence follows from Proposition C.2 and Proposition C.4.)

The following will be proved partly in class and partly on homework.

*PROPOSITION 4.5. *Let $f(x)$ be the Dirichlet function for the sequence r_k. Then $f(x)$ is continuous at α if and only if $\alpha \neq r_k$ for all k.*

In the case that r_k enumerates the rationals in $[0, 1]$, the Dirichlet function is continuous at the irrationals (which are dense) and discontinuous at the rationals (which are also dense!).

2. On a Closed Interval.

Several of the most basic and important facts of the calculus are proved for functions that are continuous on a closed interval. Notice that the statement that $f(x)$ is continuous on $[a, b]$ involves one-sided limits at the endpoints, since the definition of continuity requires us to consider $|f(x) - f(\alpha)|$ only for x, α in the domain of f.

Our first result is the *Intermediate Value Theorem*, known hence as the *IVT*. As for some results in the previous chapter, there are (at least) two proofs of this theorem. Here is a bisection proof.

**INTERMEDIATE VALUE THEOREM. *Let $f(x)$ be continuous on the closed interval $[a, b]$, and suppose that the number A is between $f(a)$ and $f(b)$. Then there is $c \in [a, b]$ such that $f(c) = A$.*

PROOF. The result is trivial if $a = b$, and so we assume that $a < b$.

We will assume that $f(a) \geq f(b)$ and leave the other case to you. We have $f(a) \geq A \geq f(b)$.

Define $a_0 = a$ and $b_0 = b$. Let $d = (a_0 + b_0)/2$. If $f(d) \geq A$, then define $a_1 = d$ and $b_1 = b_0$. If $f(d) < A$, then define $a_1 = a_0$ and $b_1 = d$. In any case,

$$f(a_1) \geq A \geq f(b_1)$$

Also observe that $b_1 - a_1 = (b_0 - a_0)/2$.

Keep going. Assume we have defined

$$a_0 \leq a_1 \leq \cdots \leq a_n < b_n \leq b_{n-1} \leq \cdots \leq b_0$$

such that $f(a_n) \geq A \geq f(b_n)$ and $b_n - a_n = (b_0 - a_0)/2^n$. Define $d = (a_n + b_n)/2$. If $f(d) \geq A$, then define $a_{n+1} = d$ and $b_{n+1} = b_n$; if $f(d) < A$, then define $a_{n+1} = a_n$ and $b_{n+1} = d$, and the pattern propagates.

We invoke Proposition 2.7 for the sequences a_n and b_n. That proposition gives them a common limit $c \in [a, b]$ such that if $\delta > 0$, then there is an integer n with $[a_n, b_n] \subset (c - \delta, c + \delta)$.

Let $\epsilon > 0$ be given, and get $\delta > 0$ for the continuity implication of $f(x)$ at c. Get n such that $[a_n, b_n] \subset (c - \delta, c + \delta)$ and then we have $|a_n - c| < \delta$ and $|b_n - c| < \delta$, and so we have

(4.2) $\qquad |f(a_n) - f(c)| < \epsilon \quad \text{and} \quad |f(b_n) - f(c)| < \epsilon$

By construction of the sequences we also have $f(b_n) \leq A \leq f(a_n)$, and so we see that

$$f(c) - \epsilon < f(b_n) \leq A \leq f(a_n) < f(c) + \epsilon$$

And so $|f(c) - A| < \epsilon$. Since ϵ is arbitrary, this proves that $f(c) = A$. ∎

An immediate corollary: the existence of all the root functions; this is an exercise at the end of the chapter. When n is a positive integer and $x \geq 0$, we write $\sqrt[n]{x}$, as expected, for the unique non-negative number whose n-th power is x. When $x < 0$ and n is an odd integer, we write $\sqrt[n]{x}$ for the unique negative number whose n-th power is x. The mere existence of the root functions does not make them continuous. Later in this chapter we will see that they are, in fact, continuous.

We will prove another basic fact: a continuous function on a closed interval has a minimum and a maximum there. It follows that a continuous function on a closed interval is bounded there; we prove that fact on its own to lead to the Extreme Value Theorem. Our proof will use an easy fact: let S and T be subsets of the domain of a function f; if f is bounded on S and on T, then it is bounded on the set $S \cup T$.

**PROPOSITION 4.6. *A continuous function on a closed interval is bounded there.*

PROOF. Let $f(x)$ be continuous on $[a, b]$. Let S be the set of $x \in [a, b]$ such that f is bounded on $[a, x]$. The function f is bounded on $[a, a]$, and so $a \in S$. Thus, S is not empty. By definition of S, it is bounded above by b. Completeness gives S a sup c.

We will prove that $c \in S$. Indeed, get $\delta > 0$ such that if $x \in [a, b]$ and $|x - c| < \delta$, then $|f(x) - f(c)| < 1$. We write the inequality $|f(x) - f(c)| < 1$ as $|f(x)| \leq |f(c)| + 1$. We see that f is bounded on $(c - \delta, c + \delta) \cap [a, b]$.

Since c is the sup of S, Proposition 1.1 finds $t \in S$ with $c - \delta < t \leq c$. Since $t \in S$, we have that f is bounded on $[a, t]$. We conclude that f is bounded on the union

$$[a, t] \cup \Big((c - \delta, c + \delta) \cap [a, b] \Big) \supseteq [a, c]$$

Thus, $c \in S$.

Next we claim that $c = b$. Assume that $c < b$, and use δ from the last paragraph. There is a number $d < b$ with $c < d < c + \delta$. The function f is bounded on $[c, d]$ by $|f(c)| + 1$. Since $c \in S$, the function f is bounded on

$[a,c]$. Thus, f is bounded on $[a,d]$, and this contradicts that c is the sup of S. Thus, $c = b$, after all.

Now $b \in S$, and so f is bounded on $[a,b]$. □

*EXTREME VALUE THEOREM. *Let $f(x)$ be continuous on the closed interval $[a,b]$. Then there are $c, d \in [a,b]$ such that $f(c) \leq f(x) \leq f(d)$ for all $x \in [a,b]$.*

PROOF. By Proposition 4.6 there is an upper bound on the image of $f(x)$, and so Completeness provides a sup α for the image. We claim that α is in the image. Suppose it is not, and define
$$g(x) = \frac{1}{\alpha - f(x)}$$
Since $\alpha - f(x)$ is not 0 on $[a,b]$, Proposition 4.3c shows that $g(x)$ is continuous on $[a,b]$. Also notice that $g(x)$ is positive there. By Proposition 4.6, there is an upper bound B to $g(x)$ on $[a,b]$. Then
$$\frac{1}{\alpha - f(x)} \leq B \quad \text{leads to} \quad f(x) \leq -\frac{1}{B} + \alpha$$
for all $x \in [a,b]$. This contradicts that α is the least upper bound of the image of f. Thus, $\alpha = f(d)$ for some $d \in [a,b]$, and we have $f(x) \leq f(d)$ for all $x \in [a,b]$.

The existence of the minimum follows similarly. □

Terminology: the minimum and maximum of a function are function values, not numbers in the domain. Thus, x^2 has maximum 9 on $[0,3]$. The maximum occurs at $x = 3$, and so we say that 3 is an *extreme point*. An *extreme point* is then a number in the domain of a function.

There is a good way to summarize the IVT and Extreme Value Theorem. This result fleshes out what we meant by saying that continuous functions are under control.

*PROPOSITION 4.7. *Let $a \leq b$ be numbers and let $f : [a,b] \to \mathbb{R}$ be continuous. Then the image of f is a closed interval.*

PROOF. The Extreme Value Theorem finds $c, d \in [a,b]$ such that $f(c) \leq f(x) \leq f(d)$ for all $x \in [a,b]$. In other words, the image of f is contained in the closed interval $[f(c), f(d)]$. The IVT says that every number between $f(c)$ and $f(d)$ is in the image as well. □

3. Inverses.

When a continuous function on an open interval is one to one, there is a *continuous* inverse function. In the following, the abbreviation IFT is used for *Inverse Function Theorem*.

*CONTINUOUS IFT. *Suppose that $f(x)$ is continuous and one to one on the open interval I. Then the image $f(I)$ is an open interval, and $f^{-1} : f(I) \to I$ is continuous.*

PROOF. We first show that $f(I)$ is an open interval. If $f(I)$ is bounded above, define β to be its sup; if it is not bounded above, let $\beta = \infty$ (and we will need to remember that β is not a number in this case). If $f(I)$ is bounded below, let α be its inf; if it is not bounded below, define $\alpha = -\infty$. We claim that $f(I) = (\alpha, \beta)$.

First, we prove that $f(I) \subseteq (\alpha, \beta)$. Let $\gamma \in f(I)$, so that $\gamma = f(b)$ for some $b \in I$. The definition of α, β shows that $\alpha \leq \gamma \leq \beta$; we claim that γ cannot be equal to either endpoint. We will give the proof that $\gamma \neq \beta$ here and prove $\gamma \neq \alpha$ in class. Assume that $\gamma = \beta$, and then β must be a number – the sup of $f(I)$. Thus, $\gamma \geq f(x)$ for all $x \in I$. Recall that $f(b) = \gamma$. Because I is an open interval, there are numbers a, c in I such that $a < b < c$. Since f is one to one, the numbers $f(a), f(b) = \gamma, f(c)$ are all distinct, and $f(b)$ is their maximum. Thus, we either have $f(a) < f(c) < f(b)$ or we have $f(c) < f(a) < f(b)$. In the first case, the IVT applied to $[a,b]$ finds $d \in [a,b]$ such that $f(d) = f(c)$. Since $b < c$, this contradicts that f is one to one. Similarly, we cannot have $f(c) < f(a) < f(b)$. We have contradicted that $\gamma = \beta$. The proof that $\gamma \neq \alpha$ then completes the proof that $f(I) \subseteq (\alpha, \beta)$.

Second, we show that $(\alpha, \beta) \subseteq f(I)$. Let $\alpha < \gamma < \beta$. If $\alpha = -\infty$, then $f(I)$ is not bounded below and we can find $a \in I$ with $f(a) < \gamma$. If α is the inf of $f(I)$, then because it is the greatest lower bound, γ is not a lower bound, and again we find $a \in I$ with $f(a) < \gamma$. A similar argument finds $b \in I$ with $\gamma < f(b)$. Since I is an open interval, all numbers between a and b are in I, and so the IVT applies to find c between a and b such that $f(c) = \gamma$. Thus, $\gamma \in f(I)$. This completes the proof that $f(I)$ is an open interval.

If $J \subseteq I$ is an open interval, then f is one to one and continuous on J, and the foregoing argument proves that $f(J)$ is an open interval, as well.

Now we prove that f^{-1} is continuous on its domain $f(I)$. Let $\gamma \in f(I)$ and let $f(x) = \gamma$. Given $\epsilon > 0$, since I is open, there is a positive number $\epsilon' \leq \epsilon$ such that $(x - \epsilon', x + \epsilon') \subseteq I$. The function f maps the open interval $(x - \epsilon', x + \epsilon')$ to an open interval (α, β) containing $f(x) = \gamma$. There is a positive number δ such that $(\gamma - \delta, \gamma + \delta) \subseteq (\alpha, \beta)$.

We claim that δ establishes the continuity implication for f^{-1} at γ. Indeed, let $|\mu - \gamma| < \delta$. Then $\mu \in (\alpha, \beta) = f(x - \epsilon', x + \epsilon')$, and so there is $c \in (x - \epsilon', x + \epsilon')$ with $f(c) = \mu$. In other words, $c = f^{-1}(\mu)$. Now $|c - x| < \epsilon'$, and this is

$$|f^{-1}(\mu) - f^{-1}(\gamma)| = |c - x| < \epsilon' \leq \epsilon$$

This proves that f^{-1} is continuous. □

It turns out that a one to one, continuous function on an open interval is either increasing or decreasing across the entire interval. This can be proved directly – the argument seems to involve an annoying number of individual cases; it is easier to see in the case that the continuous function has a derivative. We will take this up in the chapter on derivatives.

It is a straightforward consequence of the Continuous IFT that the root functions are continuous. Let P be the positive reals (an open interval!). If n is a positive integer, then the function x^n is continuous and one to one on P, and so its inverse function, the n-th root function, is continuous. It is easy to see that the image of x^n is not bounded above and that it gets arbitrarily close to 0, and since the image must be an open interval, that image must be all of P. Thus, n-th root function is continuous on the set of all positive numbers.

Once we have n-th roots, we can define rational exponents. Given integers m, n with $n > 0$ and given $x > 0$, we can define $x^{m/n} = (x^{1/n})^m$. We get all the expected properties of exponents, as you can check.

4. Uniform Continuity.

This section is motivated by what might seem a rather trivial issue of logic. In fact, when this issue is pursued, we end up with a fundamental theorem that is deeper than most of what we will do in this course. Its discovery represents a real insight into the properties of the real numbers as manifested in functions. The fact is also another example of how continuity is a strong constraint.

Let $S \subseteq \mathbb{R}$ and $f : S \to \mathbb{R}$. In the definition of continuity, we are given $\alpha \in S$ and $\epsilon > 0$ and we have to find a "delta." The number δ is allowed to depend on both α and ϵ. We want to ask whether the dependence on α is necessary.

Consider the following proof that x^2 is continuous on $[0, 1]$. Given $x, \alpha \in [0, 1]$, we notice that $|x + \alpha| \leq 2$ and we estimate

$$|x^2 - \alpha^2| = |x - \alpha| \cdot |x + \alpha| \leq |x - \alpha| \cdot 2$$

Now if we have $\epsilon > 0$ and we want $|x^2 - \alpha^2| < \epsilon$, we can let $|x - \alpha| < \epsilon/2$ and

then
$$|x^2 - \alpha^2| \le |x - \alpha| \cdot 2 < \frac{\epsilon}{2} \cdot 2 = \epsilon$$
The point here is that the one "delta" ($\epsilon/2$) works for all α.

But not always. Consider x^2 on the whole set of real numbers. Given $\epsilon > 0$, could there be a "delta" that works for all α? Given $\delta > 0$, let $x = \alpha + \delta/2$, and compute
$$\epsilon > |x^2 - \alpha^2| = |x - \alpha| \cdot |x + \alpha| = \frac{\delta}{2} \cdot \left|2\alpha + \frac{\delta}{2}\right|$$
No matter what δ is, we can choose a α large enough to violate this inequality (check that $\alpha = \epsilon/\delta$ works). So, δ cannot be chosen before we know what α is.

A definition: let $S \subseteq \mathbb{R}$ and let $f : S \to \mathbb{R}$. Then f is *uniformly continuous* on S if for all $\epsilon > 0$ there is $\delta > 0$ such that if $x, \alpha \in S$ and $|x - \alpha| < \delta$, then $|f(x) - f(\alpha)| < \epsilon$.

In the definition of uniform continuity, notice that the δ is put on the table before the α. This makes a lot of difference, as we will see. It is easy to see that a uniformly continuous function is continuous – use the one delta for all α. The fact that x^2 (on the entire set of reals) is not uniformly continuous shows that continuity does not always imply uniform continuity.

A very remarkable theorem: we always get uniform continuity when the domain of the continuous function is a closed interval. The proof is tricky, since continuity is the only available tool.

*THEOREM 4.8. *If $f(x)$ is continuous on $[a, b]$, then it is uniformly continuous there.*

PROOF. Let $\epsilon > 0$ be given. If J is a closed sub-interval of $[a, b]$ and if $\delta > 0$, we will say that δ *works for* J if we have that $x, y \in J$ with $|x - y| < \delta$ implies that $|f(x) - f(y)| < \delta$. If δ works for J, and if $0 < \mu \le \delta$, then μ works for J, as well.

Define S to be the set of $x \in [a,b]$ such that there is a $\delta > 0$ that works for $[a,x]$. We see that 1 works for $[a,a] = \{a\}$, and so $a \in S$. Let c be the sup of S. We will show that $c \in S$ and that $c = b$; this will complete the proof.

Since f is continuous at c, there is $\mu > 0$ such that if $x \in [a,b]$ and $|x - c| \leq \mu$, then $|f(x) - f(c)| < \epsilon/2$. (You will see the reason for the one-half momentarily!) Here is what we need from this: if $x, y \in [c - \mu, c + \mu]$, then

$$|f(x) - f(c)| < \epsilon/2 \quad \text{and} \quad |f(y) - f(c)| < \epsilon/2$$

and thus

$$(4.3) \qquad |f(x) - f(y)| \leq |f(x) - f(c)| + |f(c) - f(y)| < \frac{\epsilon}{2} + \frac{\epsilon}{2} = \epsilon$$

(To get ϵ in this estimate was why we used $\epsilon/2$!)

Since c is the sup of S, there is $s \in S \cap (c - \mu/2, c]$. And there is $\delta > 0$ that works for $[a,s]$. As noted above, we can decrease δ if we wish – let's have $\delta \leq \mu/2$.

Now suppose that $d \in [a,b] \cap [c, c + \mu/2]$. We claim that δ works for $[a,d]$, so that $d \in S$. Indeed, let $x, y \in [a,d]$ with $|x - y| < \delta$.

If $x, y \in [a,s]$, then the definition of δ shows that $|f(x) - f(y)| < \epsilon$. Otherwise, at least one of the two numbers is greater than s. Say $x > s$. We have

$$c - \mu/2 < s < x \leq d < c + \mu/2$$

and so $|x - c| < \mu/2$. We also have $|y - x| < \delta \leq \mu/2$, and the triangle inequality then shows that $|y - c| < \mu/2$. The estimate (4.3) then shows that $|f(x) - f(y)| < \epsilon$, and this establishes the claim.

Taking $d = c$, we see that $c \in S$. If $c < b$, then there is

$$d \in (c, c + \mu/2] \cap [a,b]$$

and the fact that $d \in S$ contradicts that c is the sup of S. $\square$

It is an exercise to show that $\sin(x)$ is uniformly continuous on the entire real numbers. However, we will see that $\sin(x^2)$ is *not* uniformly continuous on the reals. Thus, even if a continuous function is bounded, it may or may not be uniformly continuous.

When we study the integral, we will need to extend Theorem 4.8 from a single closed interval to a union of finitely many closed intervals. It is not hard to see how to do this, and it will be left as an exercise.

5. Problems

1. Let f, g be continuous on the reals, and suppose that $f(x) = g(x)$ for all rational numbers x. Show that $f(x) = g(x)$ for all real numbers x.

2. Let I be a closed interval and $f : I \to \mathbb{R}$ be continuous. Let $c \in f(I)$, and define S to be the set of $x \in I$ such that $f(x) = c$. Show that the set S has a maximum. (Hint: assume that the sup of S is not in S. Something bad happens.)

3. Let $a < b < c$ and suppose that f is continuous on $[a, b]$ and continuous on $[b, c]$. Show that f is continuous on $[a, c]$. (Hint: use the definition carefully; at b you will get two *deltas*.)

4. Let $f : \mathbb{Q} \to \mathbb{Z}$ be continuous. Prove that if $u, v \in \mathbb{Q}$ and $|f(u) - f(v)| < 1/2$, then $f(u) = f(v)$. Let S be the set of $x \in \mathbb{Q}$ such that $f(x) = 1$. Suppose that S is bounded above, and let y be the sup of S (in the real numbers). Show that $y \notin \mathbb{Q}$. (Hint: if $y \in \mathbb{Q}$, what is $f(y)$?)

5. Let $f(x)$ be the Dirichlet function for the sequence r_k, and let $a \in [0, 1]$ **not** be one of the r_k. Complete the following steps to show that $f(x)$ is continuous at a.

a) Let $\epsilon > 0$ be given. Explain why there is a positive integer n such that $1/n \le \epsilon$.

b) Explain why the set $|a - r_k|$ for $k = 1, 2, \ldots, n$ has a minimum $\delta > 0$.

c) Use δ from (b) and ϵ to get the continuity implication.

6. Prove the IVT, using the following idea: Assume that $f(a) < A < f(b)$, and let S be the set of $x \in [a, b]$ such that $f(x) \leq A$. Show that S is non-empty and bounded above. Let c be the sup of S, and prove that $f(c) = A$.

7. Prove Proposition 4.6, by contradiction using a bisection scheme. Let c be the midpoint of $[a, b]$. Assume that f is not bounded on $[a, b]$, and argue that it is either not bounded on $[a, c]$ or not bounded on $[c, b]$. Choose a sub-interval on which f is not bounded, and continue.

8. Let n be a positive integer and let α be a non-negative real number. Then there is a non-negative number β such that $\beta^n = \alpha$. (Hint: you can assume that $\alpha > 0$; let $f(x) = x^n - \alpha$ and find a closed interval $[b, c]$ such that $f(b) \leq 0$ and $f(c) \geq 0$.)

9. Show that the number β of the previous exercise is unique. (Hint: remember the induction exercise showing that x^n is an increasing function.)

10. Let n be an odd positive integer and let $\alpha \in \mathbb{R}$. Show that there is a unique $\beta \in \mathbb{R}$ such that $\beta^n = \alpha$. (Hint: you can assume that $\alpha < 0$; use the previous two exercises.)

11. Let $\alpha > 0$ and let m, n be rational numbers. Show that $(\alpha^m) \cdot (\alpha^n) = \alpha^{m+n}$. (Hint: write m, n in fraction form, and use the literal definition of $\alpha^{1/r}$ where r is an integer.)

12. Suppose that $J \subseteq I$ are closed intervals and $f : I \to J$ is continuous. Show that there is $c \in I$ such that $f(c) = c$.

13. Suppose that I is a closed interval, $f : I \to \mathbb{R}$ is continuous, and $I \subseteq f(I)$. Show that there is $c \in I$ with $f(c) = c$.

5. PROBLEMS

14. Suppose that $f : [0,1] \to \mathbb{R}$ is continuous, and there are $c_1 < c_2 < c_3$ in $[0,1]$ such that $f(c_1) = c_2$ and $f(c_2) = c_3$ and $f(c_3) = c_1$. Show that there are distinct d_1, d_2 with $f(d_1) = d_2$ and $f(d_2) = d_1$. (Note: the point c_1 has *period* 3 and d_1 has period 2. Make sure you prove that d_1, d_2 are distinct.)

15. Use the EVT to prove that $f(x) = x^6 - 6 \cdot x^5 - 3 \cdot x + 13$ has a minimum on the real numbers. (Hint: note that $f(0) = 13$. Find a positive number b such that if $|x| > b$, then $f(x) \geq 13$. Use the EVT on $[-b, b]$.)

16. Let $f(x)$ be a polynomial of even degree and with positive leading coefficient. Show that $f(x)$ has a minimum on the real numbers. (Hint: prove 16)

17. Let $S \subseteq \mathbb{R}$ and $f : S \to \mathbb{R}$. Then we say that $f(x)$ has a Δ-bound if there is a number M such that $|f(x) - f(y)| \leq M \cdot |x - y|$ for all $x, y \in S$. Show that such a function is uniformly continuous.

18. Show that $\sqrt{x}$ does not have a Δ-bound on $[0, 1]$. (Thus, not all continuous functions have a Δ-bound.)

19. (A famous alternative proof of Theorem 4.6.) Let $f : [a, b] \to \mathbb{R}$ be continuous. For each $x \in [a, b]$, find $\delta > 0$ such that f is bounded on $I_x = (x - \delta, x + \delta)$. The Heine-Borel Theorem finds a finite set F such that $[a, b]$ is the union of the I_z with $z \in F$. Show that f is bounded on $[a, b]$.

20. (A famous alternative proof of Theorem 4.8.) Let $f : [a, b] \to \mathbb{R}$ be continuous, and let $\epsilon > 0$. For each $x \in [a, b]$, find $\delta > 0$ such that if $u, v \in (x - \delta, x + \delta)$, then $|f(u) - f(v)| < \epsilon$. (Hint: think $\epsilon/2$ for x.) Define $I_x = (x - \delta/2, x + \delta/2)$. The Heine-Borel Theorem finds a finite set F such that $[a, b]$ is the union of I_z with $z \in F$. Let μ be the minimum half-width of I_z, for all $z \in F$. Show that if $x, y \in [a, b]$ and $|x - y| < \mu$, then $|f(x) - f(y)| < \epsilon$. (Hint for the last statement: $x \in I_z$ for some $z \in F$. Let $I_z = (z - \delta/2, z + \delta/2)$, and show that $|x - z| < \delta$ and $|y - z| < \delta$.)

21. Show that $f(x) = x^5 + x + 1$ is one to one without using the derivative. Show that f maps the reals *onto* the reals and so f^{-1} maps the reals onto the reals and is continuous. (Hint: show that if $a < b$, then $f(a) < f(b)$. Note: there is no algebraic formula for f^{-1}.)

22. Suppose that $f(x)$ is a decreasing continuous function on $[0, \infty)$, and assume that $f(x) \to 1$ as $x \to \infty$. Show that $f(x)$ is uniformly continuous on $[0, \infty)$. (Hint: for $\epsilon > 0$, get c such that if $x > c$, then $f(x) < 1 + \epsilon$. Use Theorem 4.8 on $[0, c]$. You will need a couple of "deltas.")

23. (This is a lemma for the next problem.) Let $p > 0$. Show that for $x, y \in \mathbb{R}$ with $|x - y| < p$, there is an integer n such that $x - n \cdot p$ and $y - n \cdot p$ are in the open interval $(0, p)$. $[-p, p]$

24. Assume that $f : \mathbb{R} \to \mathbb{R}$ is continuous. Let $p > 0$, and assume that $f(x + p) = f(x)$ for all $x \in \mathbb{R}$. (In other words, f has *period p*.) Show that f is uniformly continuous on $\mathbb{R}$. (Hint: use periodicity along with Theorem 4.8 on $[-p, p]$ and the previous problem.)

CHAPTER 5

The Derivative.

The *derivative* comes up in a multitude of contexts, especially anything applied having to do with change in time or position. You know that

$$(5.1) \qquad f'(\alpha) = \lim_{x \to \alpha} \frac{f(x) - f(\alpha)}{x - \alpha}$$

When this limit exists, we say that f is *differentiable at* α.

If T is the set of α in the domain of S such that $f'(\alpha)$ exists, then f' defines a function on T. When $T = S$ (when f' exists wherever f exists), we say that f is *differentiable*.

In applying the derivative, you have worked almost exclusively with functions defined at least on open intervals, often on the whole real line. What is the minimum required of the domain of a function to make sense of the limit definition? Since $f(\alpha)$ is referred to, the number α must be in the domain of f. Since we have $x \to \alpha$, we must have that α is a limit point of that domain as well. Thus, in order for f to be differentiable, every point of its domain must be a limit point of that domain. This is the case, for instance, when the domain is any sort of interval, except when the interval is a single point.

1. Secant Functions.

From an advanced point of view, the $x - \alpha$ in the denominator of (5.1) is trouble. The problem is that the limit notation $x \to \alpha$ automatically precludes $x = \alpha$, and this prevents us from dealing uniformly with the ratio.

Carathéodory saw how to define the derivative without using a ratio.[1] His definition is equivalent to the calculus-limit definition but easier to work with. His definition also generalizes easily to the case of functions of several variables, but we will not go into that, except to say that if you do go on to multivariate analysis, you will be prepared!

To understand Carathéodory's idea, consider the function $f(x) = x^3$, and suppose we are interested in calculating $f'(\alpha)$ for some number α. The ratio that leads to the derivative simplifies algebraically:

$$\frac{x^3 - \alpha^3}{x - \alpha} = x^2 + x \cdot \alpha + \alpha^2$$

Letting $x \to \alpha$, the right side becomes $\alpha^2 + \alpha \cdot \alpha + \alpha^2 = 3\alpha^2$, the expected derivative. We can write the factoring formula with no denominator:

$$x^3 - \alpha^3 = (x - \alpha) \cdot (x^2 + x \cdot \alpha + \alpha^2)$$

The second factor goes to the derivative $3\alpha^2$ as $x \to \alpha$. We call the function $x^2 + x \cdot \alpha + \alpha^2$ a *secant function for x^3 at α*. You might see that this function gives the slope of the secant line joining the points (α, α^3) and (x, x^3).

Here is the general definition with all the necessary technical dressing.

Definition. Suppose that S is a non-empty subset of the real numbers, and let $f : S \to \mathbb{R}$, and let $\alpha \in S$ with α a limit point of S. The function $F : S \to \mathbb{R}$ is a *secant function for f at α* if
a) $f(x) - f(\alpha) = F(x) \cdot (x - \alpha)$ for all $x \in S$;
b) $F(x)$ is continuous at α. ∎

Another example. To compute $(\sqrt{x})'$ at $x = \alpha > 0$, we compute

$$\sqrt{x} - \sqrt{\alpha} = \frac{(\sqrt{x} - \sqrt{\alpha}) \cdot (\sqrt{x} + \sqrt{\alpha})}{\sqrt{x} + \sqrt{\alpha}} = (x - \alpha) \cdot \frac{1}{\sqrt{x} + \sqrt{\alpha}}$$

---

[1] Carathéodory's idea is explained in his book: *Theory of Functions of a Complex Variable*, Chelsea, New York, 1954.

Again, the second factor, a secant function, goes to the derivative $1/(2\sqrt{\alpha})$ as $x \to \alpha$. Notice that $\sqrt{x}$ is differentiable for $x > 0$.

The continuity of $F(x)$ at α means that $\lim_{x \to \alpha} F(x) = F(\alpha)$. We will see that this value is the derivative of $f(x)$ at α.

PROPOSITION 5.1. *Suppose that S is a non-empty subset of the real numbers, and let $f : S \to \mathbb{R}$, and let $\alpha \in S$ be a limit point of S. Then $f(x)$ is differentiable at α if and only if there is a secant function for $f(x)$ at α. If $F(x)$ is a secant function for $f(x)$ at α, then $F(\alpha) = f'(\alpha)$.*

PROOF. Suppose that there is a secant function F. Since $f(x) - f(\alpha) = F(x) \cdot (x - \alpha)$, for all $x \in S$, we have
$$F(x) = \frac{f(x) - f(\alpha)}{x - \alpha} \quad \text{when} \quad \alpha \neq x \in S$$
Letting $x \to \alpha$, we keep $x \neq \alpha$. Since F is continuous, Proposition 4.1 shows that we have
$$F(\alpha) = \lim_{x \to \alpha} F(x) = \lim_{x \to \alpha} \frac{f(x) - f(\alpha)}{x - \alpha}$$
This shows that $f'(\alpha)$ exists and is $F(\alpha)$.

Suppose that f is differentiable at α. Define
$$F(x) = \begin{cases} (f(x) - f(\alpha))/(x - \alpha) & \text{if } \alpha \neq x \in S \\ f'(\alpha) & \text{if } x = \alpha \end{cases}$$
When $\alpha \neq x \in S$, the definition of F shows that $f(x) - f(\alpha) = F(x) \cdot (x - \alpha)$. When $x = \alpha$, the same equation holds, for it says that $0 = 0$. Thus,
$$f(x) - f(\alpha) = F(x) \cdot (x - \alpha) \quad \text{for all} \quad x \in S$$
We claim that F is continuous at α. Indeed, let $\epsilon > 0$ and the existence of $f'(\alpha)$ finds $\delta > 0$ such that
$$\left| f'(\alpha) - \frac{f(x) - f(\alpha)}{x - \alpha} \right| < \epsilon \quad \text{when} \quad x \in S, \ 0 < |x - \alpha| < \delta$$

For these x, the ratio is $F(x)$, and we have

$$|F(\alpha) - F(x)| < \epsilon \quad \text{when} \quad x \in S, \ 0 < |x - \alpha| < \delta$$

Since we also have $|F(\alpha) - F(\alpha)| = 0 < \epsilon$, we see that we have the continuity implication for F at α. Thus, F is continuous at α. □

Our terminology is "secant function for $f(x)$ at α," and we remind you that the secant function only gives the derivative at α – it does not give the general derivative $f'(x)$.

The secant function is continuous at the point of the derivative, but it does not need to be continuous anywhere else. In fact, the secant is continuous exactly where the function whose derivative we are taking is continuous, but we will not need this fact. We do, however, include the following fairly wild example. Define

$$f(x) = \begin{cases} x^2 & \text{if } x \in \mathbb{Q} \\ 0 & \text{if } x \notin \mathbb{Q} \end{cases}$$

$$F(x) = \begin{cases} x & \text{if } x \in \mathbb{Q} \\ 0 & \text{if } x \notin \mathbb{Q} \end{cases}$$

and observe that $f(x) = F(x) \cdot x$.

Let's make sure we understand the point of using secant functions: we want to show that they give a convenient way to establish the standard theorems about the derivative. Once we have those theorems, we can use them the way we have done in the past. Thus, to calculate the derivative of a polynomial, we will use the power rule, addition rule, etc., as you did in a calculus course. We won't use secant functions to calculate derivatives, but rather to prove theorems that allow calculation.

Here is a first example.

PROPOSITION 5.2. *If f is differentiable at α, then it is continuous at α.*

PROOF. Let S be the domain of f, and let $F(x)$ be a secant function for $f(x)$ at α. Because F is continuous at α, there is $\delta > 0$ such that if $x \in S$ and $|x - \alpha| < \delta$, then $|F(x) - F(\alpha)| < 1$. It follows that $|F(x)| < |F(\alpha)| + 1$.

Choose $\epsilon > 0$ and find a positive number μ less than ϵ and less than δ. If $x \in S$ and $|x - \alpha| < \mu$, then also $|x - \alpha| < \delta$, and we have

$$|f(x) - f(\alpha)| = |F(x) \cdot (x - \alpha)| \leq \bigl(|F(\alpha)| + 1\bigr) \cdot \mu < \bigl(|F(\alpha)| + 1\bigr) \cdot \epsilon$$

The last number is arbitrarily small, and so f is continuous at α. □

The converse can be false in a big way: let $f(x)$ be the Dirichlet function for a sequence r_k enumerating the rational numbers in $[0, 1]$. (See Chapter 4.) Then $f(x)$ does not have a derivative anywhere (although it is continuous at all the irrationals).

Back to more reasonable situations: the usual differentiation rules.

PROPOSITION 5.3. *Let $f : S \to \mathbb{R}$ and $g : S \to \mathbb{R}$ with f and g differentiable at $\alpha \in S$. Then*

***Additivity:** $f(x) + g(x)$ is differentiable at α, and*

$$(f + g)'(\alpha) = f'(\alpha) + g'(\alpha)$$

****Product Rule:** $f(x) \cdot g(x)$ is differentiable at α, and*

$$(f \cdot g)'(\alpha) = f'(\alpha) \cdot g(\alpha) + f(\alpha) \cdot g'(\alpha)$$

***Quotient Rule:** if $g(\alpha) \neq 0$, then $f(x)/g(x)$ is differentiable at α, and*

$$\left(\frac{f}{g}\right)'(\alpha) = \frac{f'(\alpha) \cdot g(\alpha) - f(\alpha) \cdot g'(\alpha)}{g(\alpha)^2}$$

PROOF. We will leave the proofs of additivity and the quotient rule to class.

Product rule: Let F be the secant function for f at α, and let G be the secant function for g at α, and compute that

$$f(x) \cdot g(x) - f(\alpha) \cdot g(\alpha) = (f(x) - f(\alpha)) \cdot g(x) + f(\alpha) \cdot (g(x) - g(\alpha))$$
$$= (x - \alpha) \cdot F(x) \cdot g(x) + f(\alpha) \cdot G(x) \cdot (x - \alpha)$$
$$= (x - \alpha) \cdot \left[F(x) \cdot g(x) + f(\alpha) \cdot G(x) \right]$$

We claim that the quantity in brackets is a secant function, for it satisfies the equation, and F, g, G are continuous at α (Proposition 5.2 applies to g). Let $x = \alpha$ in the secant and you have the product rule formula. □

*THE POWER RULE. Let n be a positive integer. Then the function x^n is differentiable at every real number, with derivative $n \cdot x^{n-1}$.

PROOF. We have the factoring formula

$$x^n - \alpha^n = (x - \alpha) \cdot \sum_{j=0}^{n-1} \alpha^j \cdot x^{n-1-j}$$

The polynomial on the right is continuous, and so it is a secant function. Its value at $x = \alpha$ gives the expected derivative. □

The Chain Rule is perhaps the most important differentiation rule. Its proof shows how easy it is to work with Caratheodory's secant functions.

**CHAIN RULE. Let D_1, D_2 be subsets of the reals. Suppose $f : D_1 \to D_2$ and $g : D_2 \to \mathbb{R}$, let f be differentiable at α and suppose that g is differentiable at $f(\alpha)$. Then $g(f(x))$ is differentiable at α, and $(g(f))'(\alpha) = g'(f(\alpha)) \cdot f'(\alpha)$.

PROOF. Let F be a secant function for f at α, and let G be a secant function for g at $f(\alpha)$. We claim that $G(f(x)) \cdot F(x)$ is a secant function for $g(f(x))$ at α. Indeed, compute for all $x \in D_1$ that

$$g(f(x)) - g(f(\alpha)) = G(f(x)) \cdot (f(x) - f(\alpha)) = G(f(x)) \cdot F(x) \cdot (x - \alpha)$$

Furthermore, F is continuous at α, and f is continuous at α and G is continuous at $f(\alpha)$. Thus, $G(f(x)) \cdot F(x)$ is continuous at α. This proves that $g(f(x))$ is differentiable at α. The derivative is

$$G(f(\alpha)) \cdot F(\alpha) = g'(f(\alpha)) \cdot f'(\alpha)$$

$\square$

You are used to using the differential notation for derivatives. If we write $y = f(x)$, imagining that x, y are variables, then we write dy/dx for $f'(x)$, a notation that we explain by saying that it is a mnemonic for things like the Chain Rule: if, to understand $g(f(x))$, we write $y = f(x)$ and $z = g(y)$, so that $z = g(f(x))$, then the Chain Rule takes the familiar form

$$\frac{dz}{dx} = \Big(g(f(x))\Big)' = g'(f(x)) \cdot f'(x) = \frac{dz}{dy} \cdot \frac{dy}{dx}$$

which looks like a statement of algebra. We will not need or use differentials formally.[2]

We use the familiar notation $f''(x)$ for the *second derivative* – the derivative of the derivative. In practice, we will not need to think about secant functions in the context of the second derivative.

2. Interior Extremes, Mean Values.

Note that the interval in the following is *open*. Also, remember that a continuous function does not have to have a maximum or minimum on an open interval – the existence of an extreme point is a *hypothesis* here.

---

[2]Differentials are most often used informally; putting them on a formal footing entails some technical gymnastics. We will stick to the informal, regarding them as a convenient shorthand. Calculus books can be confusing on this; for instance, the "definition" of the "differential" used in many books – $dy = f'(x) \cdot \Delta x$ – is not consistent with $dy/dx = f'(x)$.

**THEOREM ON INTERIOR EXTREMES. *Let I be an open interval, suppose that $f : I \to \mathbb{R}$. Suppose that f has a maximum or a minimum at $c \in I$ and that f is differentiable at c. Then $f'(c) = 0$.*

PROOF. Let E be the secant function for f at c. Suppose that $f'(c) \neq 0$, so that $E(c) \neq 0$. Since E is continuous at c, there is an open interval $J \subseteq I$ with $c \in J$ and such that $E(x)$ has the same sign as $E(c)$ for all $x \in J$. We have $f(x) - f(c) = E(x) \cdot (x - c)$, and we see that by varying the sign of $x - c$ (possible since J is *open*) we can vary the sign of $f(x) - f(c)$. This contradicts that c is an extreme point. □

It is easy to find counterexamples to this theorem if the hypothesis on I is changed at all. For instance, the function x^2 has a maximum on $(0, 1]$ where the derivative is not 0.

The Theorem on Interior Extremes is the basis for many calculus max/min problems. It might bring back happy memories to rehearse the logic. Suppose $f(x)$ is differentiable on the closed interval $[a, b]$. By the Extreme Value Theorem, $f(x)$ has a minimum on $[a, b]$; say that the minimum occurs at c. Existence is nice, but we want to find c specifically. If $a < c < b$, then the Theorem on Interior Extremes shows that $f'(c) = 0$. Otherwise $c = a$ or $c = b$. Thus, we have reduced the problem of *finding c* to the alternative equations: $f'(x) = 0$ or $x = a$ or $x = b$.

The Mean Value Theorem is the beginning of advanced level insight concerning the derivative. Notice how many facts are involved in its proof.

*MEAN VALUE THEOREM. *Let $a < b$ and suppose that $f(x)$ is continuous on $[a, b]$ and differentiable on (a, b). Then there is a number $c \in (a, b)$ such that $f(b) - f(a) = f'(c) \cdot (b - a)$.*

2. INTERIOR EXTREMES, MEAN VALUES.

PROOF. Define
$$g(x) = f(x) - f(a) - \frac{f(b) - f(a)}{b - a} \cdot (x - a)$$

We will show there is $c \in (a, b)$ such that $g'(c) = 0$. The equation $g'(c) = 0$ is the desired equation, as you should show.

Observe that $g(a) = g(b) = 0$ and that g is differentiable on (a, b). Thus, by Proposition 5.2, the function g is continuous on $[a, b]$. By the Extreme Value Theorem it has a maximum and a minimum: say the maximum is at u and the minimum at v. If u and v are endpoints of $[a, b]$, then $g(x) = 0$ for all $x \in [a, b]$ in which case $g'(x) = 0$ and so there is $c \in (a, b)$ such that $g'(c) = 0$. Otherwise, let c be whichever of u, v is not an endpoint, and the Theorem on Interior Extremes says that $g'(c) = 0$. □

Notice that if $a = b$ in the hypothesis of the Mean Value Theorem, the equation of the conclusion reads $f(b) - f(b) = f'(c) \cdot (b - b)$, which says that $0 = 0$. Thus, we can assume that $a \leq b$ rather than $a < b$ in the Mean Value Theorem. Furthermore, suppose that $a > b$, and then the equation in the Mean Value Theorem is this: $f(a) - f(b) = f'(c) \cdot (a - b)$. Multiply by -1 to obtain $f(b) - f(a) = f'(c)(b - a)$, the form of the equation on the conclusion of the Mean Value Theorem. In other words, the equation of the theorem holds when $a > b$ (although the "c" should be described as being *between* a and b and not in the open interval (a, b)). Thus, the hypothesis of the Mean Value Theorem can be simply that $f(x)$ is continuous between and including a and b and differentiable between them. We will feel free to use this more general version of the theorem.

Many of the exercises at the end of this section give standard uses of the Mean Value Theorem. For example, we establish the well-known link between the sign of the derivative and the oscillation of its graph. We need some definitions. Let I be an interval and $f : I \to \mathbb{R}$. We say that f is *increasing* if $f(a) \leq f(b)$ when $a \leq b$ and $a, b \in I$. We say that f is *strictly increasing* if $f(a) < f(b)$ when $a < b$ and $a, b \in I$. Similar to the use for sequences, *increasing* involves non-strict inequalities. The definitions of *decreasing* and *strictly decreasing* are left to you. In the exercises you will link *increasing* and *decreasing* to the expected signs of $f'(x)$.

3. Inverse Functions.

The Continuous IFT in Chapter 4 showed us that the inverse of a continuous function on an open interval is continuous. Now we show that the inverse is differentiable, provided that the original function has a non-zero derivative. The final equation in the conclusion of this result needs to be studied carefully.

*DIFFERENTIABLE IFT. Let D be an open interval on which f is differentiable. Suppose that $f'(x) \neq 0$ for all $x \in D$. Then f is one to one on D, and f^{-1} is differentiable on $f(D)$. Furthermore, if $b \in D$ and $\beta = f(b)$, then $f'(b) \cdot (f^{-1})'(\beta) = 1$.

PROOF. To see that f is one to one, suppose that $u, v \in D$ are distinct. Use the Mean Value Theorem to find z between u, v such that $f(u) - f(v) = f'(z) \cdot (u - v)$. Since $u \neq v$ and $f'(z) \neq 0$, it must be that $f(u) \neq f(v)$, and this proves that f is one to one.

Since $f(x)$ is one to one and continuous, the Continuous IFT shows that the set $f(D)$ must be an open interval and that $f^{-1} : f(D) \to D$ is continuous as well. Let $b \in D$, and let $\beta = f(b)$, so that $f^{-1}(\beta) = b$. Let F be the secant function for f at b, and we claim that $F(x) \neq 0$ for all x. Indeed, $f(x) - f(b) = F(x) \cdot (x - b)$. If $x \neq b$, then since f is one to one, we have $f(x) \neq f(b)$, and we see that $F(x) \neq 0$. Also, $F(b) = f'(b) \neq 0$.

Since F is not 0, the function $1/F(f^{-1}(y))$ is defined for all $y \in f(D)$, for the number $f^{-1}(y)$ is an element of the domain D of f. We claim that this function is a secant function for f^{-1} at β. Indeed,

$$y - \beta = f(f^{-1}(y)) - f(f^{-1}(\beta)) = F(f^{-1}(y)) \cdot (f^{-1}(y) - f^{-1}(\beta))$$

Since $F \neq 0$, we can divide by it to obtain

$$\frac{1}{F(f^{-1}(y))} \cdot (y - \beta) = f^{-1}(y) - f^{-1}(\beta)$$

for all $y \in f(D)$. This proves that $1/F(f^{-1}(y))$ satisfies the necessary secant equation. Since f^{-1} is continuous and since F is continuous at b, Proposition 4.4 and Proposition 4.3c show that $1/F(f^{-1}(y))$ is continuous at β. The differentiability of f^{-1} follows as does the equation relating the derivatives of f and f^{-1}. $\square$

There is a symbolic differential form of the IFT: if we write $y = f(x)$, thinking of x, y as variables, and then write $f^{-1}(y) = x$, the concluding equation is that

$$f'(x) \cdot (f^{-1})'(y) = 1 \quad \text{which is} \quad \frac{dy}{dx} \cdot \frac{dx}{dy} = 1$$

which latter equation looks like obvious algebra.

An exercise shows that the Differentiable IFT gives the Power Rule for *rational number exponents*. Notice that x^3 has a *continuous* inverse at all real numbers, the cube root function is not differentiable at 0, where the derivative of x^3 is 0.

Here is another common use of the Differentiable IFT. Suppose that the function $\sin(\theta)$ has been defined, and suppose we know that it has derivative $\cos(\theta)$ and that $\cos(\theta) > 0$ for $-\pi/2 < \theta < \pi/2$. Suppose also that we know the Pythagorean identity $\sin^2(\theta) + \cos^2(\theta) = 1$. The Differentiable IFT defines an inverse function $\arcsin(x)$ for $\sin(x)$, where if $x = \sin(\theta)$, then $\arcsin(x) = \theta$ and

$$\arcsin'(x) = \frac{1}{\sin'(\theta)} = \frac{1}{\sqrt{1 - \sin^2(\theta)}} = \frac{1}{\sqrt{1 - x^2}}$$

Remember that θ is in the open interval $(-\pi/2, \pi/2)$. The same idea finds the function $\arctan(x)$ and its derivative $1/(1+x^2)$.

4. L'Hôpital's Rule.

We will almost certainly not have time to discuss these proofs in class. They are given here because it is unlikely that you have seen rigorous arguments for these important facts.

We need a generalization of the Mean Value Theorem.

CAUCHY'S MEAN VALUE THEOREM. *Let $f(t)$ and $g(t)$ be differentiable for t between the two distinct real numbers a, b. Suppose that g' is non-zero between a, b. Then there is c between a, b such that*

$$\frac{f(b) - f(a)}{g(b) - g(a)} = \frac{f'(c)}{g'(c)}$$

PROOF. Define

$$h(t) = f(t) - f(a) - (g(t) - g(a)) \cdot \frac{f(b) - f(a)}{g(b) - g(a)}$$

and observe that $h(a) = 0 = h(b)$. The Mean Value Theorem finds c between a, b such that $h'(c) = 0$, and you can compute that this is the desired equation. (Note that $g(b) \neq g(a)$. Why?) □

Suppose we have an open interval I in the reals. Let α be an endpoint of I (so that α could be real or ∞ or $-\infty$). Suppose that $f(x)$ is differentiable in I, that $g(x)$ has non-zero derivative in I, and that L is a real number with

(5.2) $$\lim_{x \to \alpha} \frac{f'(x)}{g'(x)} = L$$

Because g has non-zero derivative in I, the Mean Value Theorem shows that there is at most one $c \in I$ such that $g(c) = 0$. We can make I smaller, if necessary, so that $g(x) \neq 0$ on I.

4. L'HÔPITAL'S RULE.

Let $\epsilon > 0$. Because of the limit (5.2), we can replace I by a smaller open interval, if necessary, so that

(5.3) $$\left|\frac{f'(c)}{g'(c)} - L\right| < \epsilon \quad \text{for all} \quad c \in I$$

Let a and b be distinct elements of I. We can apply Cauchy's Mean Value Theorem to $g(t), f(t)$ over the interval between $t = a$ and $t = b$, to conclude that there is c between a and b such that

$$\frac{f(b) - f(a)}{g(b) - g(a)} = \frac{f'(c)}{g'(c)}$$

Since $g(b) \neq 0$, we can solve for $f(b)/g(b)$:

(5.4) $$\frac{f(b)}{g(b)} = \frac{f(a)}{g(b)} + \left(1 - \frac{g(a)}{g(b)}\right) \cdot \frac{f'(c)}{g'(c)}$$

This equation holds for $a = b$, too, and so we will consider (5.4) for all $a, b \in I$.

Here are two common cases of L'Hôpital's Rule.

Case 0/0 Assume that $f(x)$ and $g(x)$ go to 0 as $x \to \alpha$. Letting $a \to \alpha$ in (5.4) and using (5.3) for each c encountered along the way, we see that

(5.5) $$\left|\frac{f(b)}{g(b)} - L\right| \leq \epsilon$$

This holds for all $b \in I$, and since ϵ is arbitrary, we see that

(5.6) $$\lim_{x \to \alpha} \frac{f(x)}{g(x)} = L$$

What is common called the "∞/∞" case really only needs that the denominator goes to infinity.

Case $/\infty$ Assume that $|g(x)| \to \infty$ as $x \to \alpha$. Fixing a in (5.4) and remembering (5.3), we can get b close enough to α so that (5.5) holds. Once again, we get (5.6).

You are familiar with some of the standard limits that follow from L'Hôpital's Rule. Here is an assortment.

$$\lim_{x\to\infty} \frac{x^n}{e^x} = 0 \quad \text{for every rational } n$$

$$\lim_{x\to\infty} \frac{\ln(x)}{x^n} = 0 \quad \text{for every positive rational } n$$

$$\lim_{x\to 0} x^n \cdot \ln(x) = 0 \quad \text{for every positive number } n$$

$$\lim_{x\to 0} \frac{\sin(x)}{x} = 1$$

5. Problems

1. Find a secant function for $(x^2+1)/(3x-1)$ at 4. (Hint: write the difference involved as one fraction and look for $(x-4)$.)

2. Look up the Taylor series formula for e^x and use it to find a series formula for the secant function for e^x at 0. (You may assume that the power series is continuous.)

3. Prove that if $f(x)$ is differentiable at α and if $c \in \mathbb{R}$, then $c \cdot f(x)$ is differentiable at α, and $(c \cdot f(x))'(\alpha) = c \cdot f'(\alpha)$.

4. Let Q be the set of positive rational numbers and define $q : Q \to \mathbb{R}$ by $q(x) = 2^x$. Assume that the following limit exists.[3]

$$\lim_{x\to 0} \frac{2^x - 1}{x}$$

Show that 2^x is differentiable on Q and find a formula for its derivative. (Remember that x is a *rational* number in this problem.)

5. Show that $(x^5 + 5 \cdot x^4 + 20 \cdot x^3 + 60 \cdot x^2 + 121 \cdot x + 122) \cdot e^{-x}$ has a maximum on the real numbers, and find the extreme point where the maximum occurs.

---

[3]Later in the course, we will show that this limit is, in fact, $\ln(2)$; for now just assume that it exists.

6. (Pasting functions together.) Let $a < b < c$. Suppose that $f : [a, b] \to \mathbb{R}$ with $f'(x)$ continuous, and suppose that $g : [b, c] \to \mathbb{R}$ with $g'(x)$ continuous. Suppose that $f(b) = g(b)$ and $f'(b) = g'(b)$. Show that there is $h : [a, c] \to \mathbb{R}$ with $h'(x)$ continuous and $h = f$ on $[a, b]$ and $h = g$ on $[b, c]$.

7. Prove the following statements. In each of them, I is an interval in the real numbers and f, g are functions on I.
  a) Suppose that $f'(x) = 0$ for all $x \in I$. Show that f is constant on I.
  b) Suppose that $f'(x) = g'(x)$ on I. Then there is a constant c such that $f(x) = g(x) + c$ on I.
  c) If $f'(x) \geq 0$ on I, then $f(x)$ is increasing on I.
  d) If $f'(x) > 0$ on I, then $f(x)$ is strictly increasing on I.

8. Assume the following things (and only these) about the trigonometric functions.
  1) $\sin(x)$ and $\cos(x)$ are defined on $\mathbb{R}$;
  2) $\cos(0) = 1$ and $\sin(0) = 0$;
  3) $\cos'(x) = -\sin(x)$ and $\sin'(x) = \cos(x)$.
  Prove that $\cos^2(x) + \sin^2(x) = 1$. (Hint: take the derivative of the left side.)

9. Make the same assumptions as in the previous problem. Suppose that $f : \mathbb{R} \to \mathbb{R}$ and that $f''(x) = -f(x)$ for all $x \in \mathbb{R}$.
  a) Show that $f(x) \cdot \cos(x) - f'(x) \cdot \sin(x)$ has derivative 0.
  b) Show that the function in (a) is equal to $f(0)$.
  c) Show that $f'(x) \cdot \cos(x) + f(x) \cdot \sin(x)$ has derivative 0.
  d) Show that the function in (c) is $f'(0)$.
  e) Solve equations (b) and (d) to show that
  $$f(x) = f(0) \cdot \cos(x) + f'(0) \cdot \sin(x)$$

10. Show that the previous problem can be applied to the function $\sin(x+a)$ where a is a constant. Deduce the angle addition formula for $\sin(x)$.

11. Show that $\sin(x^2)$ is not uniformly continuous on $[0, \infty)$. (Hint: take $\epsilon = 1$; let x, y be very close to $\sqrt{k \cdot \pi}$ for a positive integer k; use the MVT to show that $|\sin(x^2) - \sin(y^2)|$ can be large.)

12. Assume that $\exp : \mathbb{R} \to \mathbb{R}$, that $\exp(x)$ is its own derivative and that $\exp(0) = 1$. Do not assume anything else about this function.

a) Show that $\exp(x) \cdot \exp(-x) = 1$. (Hint: take the derivative!)

b) Let $a \in \mathbb{R}$. Show that $\exp(x + a) \cdot \exp(-x) = \exp(a)$.

c) Show that $\exp(x + a) = \exp(x) \cdot \exp(a)$.

13. Let $f : \mathbb{R} \to \mathbb{R}$ with $f'(x)$ continuous. Let $p \in \mathbb{R}$ and suppose that $f(p) = p$ and that $|f'(p)| < 1$.

a) Choose r with $|f'(p)| < r < 1$. Use the continuity of $f'(x)$ to find numbers $a < p < b$ such that if $x \in [a, b]$, then $|f'(x)| \leq r$.

b) Define $x_0 = a$ and $x_{n+1} = f(x_n)$ for $n = 0, 1, 2, \ldots$. Show that $|x_n - p| \leq |x_0 - p| \cdot r^n$ for all n (Hint: induction. Remember that $x_{n+1} = f(x_n)$; use the MVT.)

c) Show that $x_n \to p$ as $n \to \infty$.

(Note: this problem contains a very commonly used technique for approximating the solution of equations.)

14. Let $f(x)$ be differentiable on $[0, 1]$, and suppose that $f(0) = 0$ and $f(1) = 0$. Show that there is x with $0 < x < 1$ and $f(x) = f'(x)$. (Hint: think about $f(x) \cdot e^{-x}$ on $[0, 1]$.)

15. Use the formula for the derivative of x^n when n is a positive integer to prove the formula for the derivative of $x^{m/n}$ where m, n are positive integers.

5. PROBLEMS

16. Suppose that $f(x)$ has a continuous derivative[4] on $[a,b]$. Then $f(x)$ has a Δ-bound. (Recall that it is easy to show that a function with a Δ-bound is uniformly continuous. Thus, it is easy to prove the uniform continuity of *differentiable* functions on a closed interval.)

17. Follow the steps given to prove the *Differentiable IVT*: Suppose that $f : [a,b] \to \mathbb{R}$ is differentiable, and suppose that $f'(a) < C < f'(b)$ for some number C. Then there is $c \in [a,b]$ such that $f'(c) = C$. (The same conclusion holds if $f'(a) > C > f'(b)$.)

  (1) The number $(f(a) - f(b))/(a - b)$ is either less then or equal to C, or it is greater than C. Suppose that it is less than or equal to C. (We will be content with this case, but you should think about the other one.) Then there is $b_1 \in (a,b)$ such that
$$C < \frac{f(b_1) - f(b)}{b_1 - b}$$
  (2) The function $g(x) = (f(x) - f(b))/(x - b)$ is continuous on $[a, b_1]$, and we have $g(a) \leq C < g(b_1)$. Thus, there is $c \in [a, b_1]$ such that $g(c) = C$.
  (3) The MVT completes the proof!

---

[4] Functions with a continuous derivative are called C^1 functions; they are extremely prominent in applied mathematics.

CHAPTER 6

The Definite Integral.

1. Introduction to Integration.

You have been introduced to the integral through two ideas, combined in the Fundamental Theorem of Calculus: if $f(x)$ is continuous, then

(A) $$\int_a^b f(x) \cdot dx = F(b) - F(a) \quad \text{where} \quad F'(x) = f(x)$$

and

(B) $$\int_a^b f(x) \cdot dx = \lim \sum_i f(x_i) \Delta x_i$$

where the x_i are chosen inside subintervals of length Δx_i which divide up the interval $[a, b]$. Equation A is used primarily to compute the values of integrals, and equation B is used to recognize applications – that the integral computes some desired quantity which is approximated by the Riemann sums[1] on the right hand side.

Up until the early nineteenth century, the connection between equations A and B was treated informally, with a certain element of vagueness concerning the precise nature of the limit in equation B. This vagueness did not pose a problem; on the contrary, Newton, Leibnitz, Euler and many others used the integral to obtain concrete formulas and theoretical implications that solved longstanding problems in the physical sciences, engineering, geometry, number theory, the representation and properties of functions, etc. Equation A

---

[1]Sums like those in (B) were considered before the name *Riemann sum* was applied. Strange but true.

seemed to be the main fact, and it was used even (in fact especially!) when the integrand $f(x)$ was a power series. Everything seemed to work perfectly, despite the marginal discomfort felt by some regarding the informal nature of calculations coming from equation B.

But through the work of Euler, the Bernoulli's, and Fourier on infinite series of trigonometric functions, and that of Cauchy on power series, the defects of the informal approach became more and more apparent. It was clear what was needed: a more formal definition of the integral and a rigorous verification of its properties. Of the various definitions that were suggested during the nineteenth century, that of Riemann seemed to be best, therefore the integral that one uses in calculus is called the *Riemann integral*. It may seem strange to you that there would be any latitude at all in defining the integral, and in a sense there isn't: we certainly want it to be the case that equation A holds. What is more subtle is that equation A does not tell the whole story, and so it is not a good starting place. We will certainly *end up* there when the integrand is continuous!

Our own treatment of this subject has been strongly influenced by two ideas. First, we find the subject easier to understand if we separate the problem of the existence of the integral from the problem of computing its value. The functions for which the integral exists are called "integrable," and it is possible to define "integrable" in a way that does not mention the integral directly. The advantage of this approach is to focus on a specific type of sum (the "variation sum") before diving into integral calculations and properties. Second, imitating an idea of Darboux, we define the integral as the unique number sandwiched between certain Riemann sums, rather than as a limit of arbitrary Riemann sums. This approach makes it easier to understand and work with the integral.

2. Variation Sums and Integrable Functions.

We will need to study what is called the "variation" of a function. To define this, let $f(x)$ be bounded on $I \subseteq \mathbb{R}$. If $|f(x)| \leq A$ for all $x \in I$, then it is easy to estimate, for $x, y \in I$,

$$|f(x) - f(y)| \leq |f(x)| + |f(y)| \leq 2A$$

Thus, the set of $|f(x) - f(y)|$ for all $x, y \in I$ is bounded above. The sup of this set is called the *variation of f on I* and is denoted $\mathrm{var}(f, I)$. It is not hard to see that if I is a closed interval and $f(x)$ is continuous on I, then $\mathrm{var}(f, I)$ is the difference between the maximum value of f and the minimum value of f.

We will need the following alternative formula for the variation.

PROPOSITION 6.1. *Let $f(x)$ be bounded on $I \subseteq \mathbb{R}$, and let s be the sup of $f(I)$ and i the inf of $f(I)$. Then $\mathrm{var}(f, I) = s - i$.*

PROOF. Let $x, y \in I$ and we have $f(x) \leq s$ and $f(y) \geq i$. Multiplying the second inequality by -1 we have $-f(y) \leq -i$. Adding this to $f(x) \leq s$, we see that $f(x) - f(y) \leq s - i$. Trading places: $f(y) \leq s$ and $-f(x) \leq -i$, so that $f(y) - f(x) \leq s - i$.

The number $|f(x) - f(y)|$ is equal to one of the numbers: $f(x) - f(y)$ and $f(y) - f(x)$. We have that $|f(x) - f(y)| \leq s - i$. That this is true for all $s, y \in I$ proves that $\mathrm{var}(f, I) \leq s - i$.

Let $\epsilon > 0$. The definition of s as sup finds $x \in I$ such that $s - \epsilon < f(x)$. Similarly there is $y \in I$ such that $f(y) < i + \epsilon$. Multiplying this by -1 we have $-i - \epsilon < -f(y)$. And we have

$$s - \epsilon - i - \epsilon < f(x) - f(y) \leq |f(x) - f(y)| \leq \mathrm{var}(f, I)$$

We see that $s - i < \mathrm{var}(f, I) + 2\epsilon$ for all $\epsilon > 0$. It follows that $s - i \leq \mathrm{var}(f, I)$ and we have equality. $\square$

Next we attempt to confuse you by defining two terms for the same thing: subdivision and partition. Both of these terms refer to chopping up a closed interval into a finite number of closed sub-intervals. Given numbers $a \leq b$, a *subdivision* of $[a,b]$ is a finite set of points in the interval, with the stipulation that the endpoints of the interval must be included. The elements of such a set can always be written in order $a = x_0 < x_1 < \ldots < x_n = b$. The *partition* P associated with this subdivision is the finite set of intervals $P = \{\,[x_0, x_1], [x_1, x_2], \ldots, [x_{n-1}, x_n]\,\}$. Usually it will be more convenient to work with the partition than the subdivision, but whenever we say, "let P be a partition," we understand that the partition comes from a subdivision. It is easy to see that the sum of the lengths of the intervals in a partition of $[a,b]$ is the length $b - a$ of the whole interval. In general, we will denote the length of a closed interval I by $|I|$, and so we have

$$\sum_{I \in P} |I| = b - a$$

In the case that $a = b$, we allow just one partition: $\{[a,a]\}$. We need this case to avoid fussing when it comes up in practice; it is the only case where the closed interval or closed sub-interval is allowed to contain only one point.

Let $f(x)$ be bounded on $[a,b]$, and let P be a partition of $[a,b]$. Define the *variation sum for f on P*

$$\Sigma_l^u(f, P) = \sum_{I \in P} \operatorname{var}(f, I) \cdot |I|$$

In class we will draw a picture of this sum – it looks like the sum of the areas of rectangles, one rectangle for each sub-interval I in the partition P, that "follow along the curve $y = f(x)$."

We mentioned before that when $a = b$ we allow $P = \{[a,a]\}$. Notice in this case that $\Sigma_l^u(f, P) = 0$ no matter what f is.

Suppose that $f(x)$ is bounded on $[a,b]$. Then we say that f is *integrable on $[a,b]$* if there are arbitrarily small variation sums. In other words, for every

$\epsilon > 0$, there is a partition P of $[a,b]$, such that $\Sigma_l^u(f,P) < \epsilon$. Later, we will see that the integral $\int_a^b f$ is defined when f is an integrable function; as we mentioned before, it turns out to be a good idea to *avoid* considering the value of the integral at first.

We want to identify a large class of integrable functions, although we will stop short of a full characterization of such functions.

PROPOSITION 6.2. *Let $f(x)$ be continuous on $[a,b]$. Then $f(x)$ is integrable on $[a,b]$.*

PROOF. Continuous functions are bounded, and so $f(x)$ is bounded on $[a,b]$. Let $\epsilon > 0$. Theorem 4.8 tells us that a continuous function on a closed interval is uniformly continuous there, and so there is $\delta > 0$ such that $|x-y| < \delta$ for $x,y \in [a,b]$ implies that $|f(x) - f(y)| < \epsilon$.

Suppose that I is a closed interval contained in $[a,b]$ with $|I| < \delta$. If $x,y \in I$, then $|x-y| < \delta$, so that $|f(x) - f(y)| < \epsilon$. This shows that $\mathrm{var}(f,I) \leq \epsilon$.

Let P be a partition of $[a,b]$ such that each $I \in P$ has $|I| < \delta$. Then

$$\Sigma_l^u(f,P) = \sum_{I \in P} \mathrm{var}(f,I) \cdot |I| \leq \sum_{I \in P} \epsilon \cdot |I| = \epsilon \cdot (b-a)$$

The number $\epsilon \cdot (b-a)$ can be made arbitrarily small. □

We point out some obvious consequences of Proposition 6.2. Every polynomial is integrable over every closed interval. Every rational or algebraic function is integrable over every closed interval inside its domain. Every function differentiable on a closed interval is integrable there.

We can generalize Proposition 6.2 to include some discontinuities.

*PROPOSITION 6.3. *Let $f : [a,b] \to \mathbb{R}$ be bounded and continuous except possibly at finitely many points. Then $f(x)$ is integrable on $[a,b]$.*

6. THE DEFINITE INTEGRAL.

PROOF. The proof is like that of Proposition 6.2 except that we have to step around the discontinuities. The finitely many discontinuities can be written in numerical order; it will be convenient to include a, b along with them, and so we can get

$$a = a_0 < a_1 < \cdots < a_n = b$$

where if $c \in [a, b]$ is not one of the a_i's, then $f(x)$ is continuous at c. Also, remember that n is constant throughout the proof.

Let $\epsilon > 0$ be given. Choose $t > 0$ smaller than ϵ and smaller than the numbers $(a_i - a_{i-1})/2$ (for $1 \leq i \leq n$). Define $A_0 = [a_0, a_0 + t]$, define $A_i = [a_i - t, a_i + t]$ for $1 \leq i \leq n - 1$, and define $A_n = [a_n - t, a_n]$. The intervals A_i do not overlap.

We will construct a partition using the A_i and other intervals. To get the other intervals, consider what happens when you take the A_i out of $[a, b]$. We get a union of open intervals. Since partitions need closed intervals, we include the endpoints of these intervals. Here is what we get: define

$$D = [a_0 + t, a_1 - t] \cup [a_1 + t, a_2 - t] \cup$$
$$\cdots \cup [a_{n-2} + t, a_{n-1} - t] \cup [a_{n-1} + t, a_n - t]$$

If $c \in D$, then c cannot be one of the a_i, and so f is continuous at c. Since D is a finite union of closed intervals, it follows that f is uniformly continuous on D. (This generalization of Theorem 4.8 was mentioned after the proof of that theorem.)

Recall that $\epsilon > 0$ is given. Choose $\delta > 0$ such that if $x, y \in D$ and $|y - x| < \delta$, then $|f(x) - f(y)| < \epsilon$. Thus, if I is a closed interval contained in D, and $|I| < \delta$, then $\text{var}(f, I) \leq \epsilon$.

Let P be a partition of $[a, b]$ using the A_i with $0 \leq i \leq n$ as well as other intervals (each contained in D) of width less than δ. Let G name these other intervals. Thus if $I \in P$ then either $I = A_i$, so that $|I| \leq 2t \leq 2\epsilon$, or $I \in G$, so that $\mathrm{var}(f, I) \leq \epsilon$.

Let $B = \mathrm{var}(f, [a, b])$. If I is a subinterval of $[a, b]$, then we know that $\mathrm{var}(f, I) \leq B$. Compute

$$\Sigma_l^u(f, P) = \sum_{I \in P} \mathrm{var}(f, I) \cdot |I| = \sum_{i=0}^{n} \mathrm{var}(f, A_i) \cdot 2t \; + \; \sum_{I \in G} \mathrm{var}(f, I) \cdot |I|$$

$$\leq n \cdot B \cdot 2 \cdot \epsilon + \epsilon \cdot (b - a) = \epsilon \cdot (2Bn + b - a)$$

Because a, b, B, n are constants, this last expression can be made arbitrarily small. □

Proposition 6.3 applies to such discontinuous functions as

$$f(x) = \begin{cases} \sin(x^{-1}) & \text{when } x \neq 0 \\ 0 & \text{when } x = 0 \end{cases}$$

The Dirichlet functions defined in Chapter 4 furnish a class of even crazier examples. Recall how these functions were constructed: from a closed interval $[a, b]$, choose a one to one sequence c_k for $k \geq 1$. Define $f(c_k) = 1/k$, and define $f(x) = 0$ if x is not one of the c_k. In class we will show that the Dirichlet functions are integrable.

The Dirichlet functions are discontinuous at the infinitely many c_k. This shows that an integrable function can be badly discontinuous. On the other hand, define $g(x)$ on $[0, 1]$ to be 1 if x is rational and 0 if x is irrational. An exercise shows that $g(x)$ is not integrable. The two examples: the Dirichlet functions and the present $g(x)$, show that the case of infinitely many discontinuities is ambiguous.

In our next result continuity is not mentioned at all. Proposition 6.4 mentions increasing functions; there is an analogous result for decreasing functions – you are invited to provide the proof.

Proposition 6.4. *Let $f(x)$ be defined and increasing on $[a, b]$. Then $f(x)$ is integrable on $[a, b]$.*

Proof. We see that $f(a)$ and $f(b)$ are bounds on $f(x)$, and so f is bounded. Given a sub-interval $[c, d]$ contained in $[a, b]$, observe that $x, y \in [c, d]$ implies that $f(x) \leq f(d)$ and $f(y) \geq f(c)$, so that $f(x) - f(y) \leq f(d) - f(c)$. Since $f(c) \leq f(d)$, we have

$$f(d) - f(c) = |f(d) - f(c)|$$

and this shows that $f(d) - f(c) = \text{var}(f, [c, d])$.

Let P be the partition $[a, b]$ coming from a sub-division x_j (for $0 \leq j \leq n$) with the x_j equally spaced. Each interval in P has width $(b-a)/n$. In light of all this, the variation sum collapses:

$$\Sigma_l^u(f, P) = \sum_{j=1}^{n}(f(x_j) - f(x_{j-1}))\frac{b-a}{n} = (f(b) - f(a))\frac{b-a}{n}$$

This last term can be made arbitrarily small by making n large. $\square$

Proposition 6.4 can be generalized easily to the situation where the closed interval is a finite union of intervals on which the function is either increasing or decreasing. For instance, polynomials have this property – this gives us an additional way of seeing that polynomials are integrable. The advantage of Proposition 6.4 is that it shows specifically how to chop up the interval to make the variation sum small. This can be useful in approximating integrals.

3. Algebraic Combinations of Integrable Functions.

Here is a simple fact that we will use repeatedly.

Proposition 6.5. *Let $f(x)$ be bounded on the set J and suppose that $I \subseteq J$. Then $\text{var}(f, I) \leq \text{var}(f, J)$.*

PROOF. If $x, y \in I$, then $x, y \in J$, so that $|f(x) - f(y)| \leq \mathrm{var}(f, J)$. Because $x, y \in I$ were arbitrary, this proves that $\mathrm{var}(f, J)$ is an upper bound for the set on which $\mathrm{var}(f, I)$ is the least upper bound. Thus, $\mathrm{var}(f, I) \leq \mathrm{var}(f, J)$. □

The set of all partitions of a closed interval is hard to picture. Fortunately, we can confine our study to pairs of partitions that are related in a very specific way. We say that the partition R *refines* the partition P if each subdivision in P is a subdivision in R. In other words, suppose that the subdivision of P is
$$a = x_0 < x_1 < \ldots < x_n = b$$
Then the x_i are part of (maybe all of) the subdivision of R. Refining a partition cannot increase the variation sum.

*PROPOSITION 6.6. *Let $f : [a, b] \to \mathbb{R}$ be bounded. If P and R are partitions of $[a, b]$, and if R refines P, then $\Sigma_l^u(f, R) \leq \Sigma_l^u(f, P)$*

PROOF. We need notation to relate R and P. Given one of the intervals $I \in P$, the intervals in R that are contained in I make up a partition of I. Let I_R be this set of intervals. To repeat, I_R is the set of all $J \in R$ such that $J \subseteq I$. The sum of $|J|$ over all $J \in I_R$ is $|I|$. Furthermore, each $J \in R$ is in I_R for a unique element I of P. (Note that this is because we do not allow a closed interval in a partition to consist merely of one point!) In light of this, we see that

(6.1) $$\Sigma_l^u(f, R) = \sum_{J \in R} \mathrm{var}(f, J) \cdot |J| = \sum_{I \in P} \sum_{J \in I_R} \mathrm{var}(f, J) \cdot |J|$$

For $I \in P$ and $J \in I_R$ we have $J \subseteq I$, and so by Proposition 6.5 we have $\mathrm{var}(f, J) \leq \mathrm{var}(f, I)$. We use this to continue the calculation (6.1):

(6.2) $$\sum_{I \in P} \sum_{J \in I_R} \mathrm{var}(f, J) \cdot |J| \leq \sum_{I \in P} \sum_{J \in I_R} \mathrm{var}(f, I) \cdot |J|$$

Since $\operatorname{var}(f, I)$ does not depend on J, we can factor it out:
$$(6.3) \qquad \sum_{I \in P} \sum_{J \in I_R} \operatorname{var}(f, I) \cdot |J| = \sum_{I \in P} \operatorname{var}(f, I) \cdot \sum_{J \in I_R} |J|$$
The sum of $|J|$ over $J \in I_R$ is $|I|$, and so
$$\sum_{I \in P} \operatorname{var}(f, I) \cdot \sum_{J \in I_R} |J| = \sum_{I \in P} \operatorname{var}(f, I) \cdot |I| = \Sigma_l^u(f, P)$$
Putting (6.1) and (6.2) and (6.3) together with this last equality, we obtain $\Sigma_l^u(f, R) \leq \Sigma_l^u(f, P)$. $\square$

We will need the fact that if P and Q are given partitions, then there is a partition R that refines both of them. This is obvious: for instance, let the sub-division for R be the union of the sub-divisions for P and Q.

Next we show that sums and products of integrable functions are integrable. In the case of the sum, you know that the integral of a sum is the sum of integrals; that result will be proved later.

*PROPOSITION 6.7. *Let $f(x)$ and $g(x)$ be integrable on $[a, b]$. Then the functions $f(x) + g(x)$ and $f(x) \cdot g(x)$ are integrable there.*

PROOF. Choose $\epsilon > 0$. Since $f(x)$ is integrable, there is a partition P such that $\Sigma_l^u(f, P) < \epsilon$, and similarly that g is integrable produces a partition Q such that $\Sigma_l^u(g, Q) < \epsilon$. Let R be a partition that refines both P and Q. By Proposition 6.6, we have $\Sigma_l^u(f, R) \leq \Sigma_l^u(f, P)$ and that $\Sigma_l^u(g, R) \leq \Sigma_l^u(g, Q)$. Thus $\Sigma_l^u(f, R) < \epsilon$ and $\Sigma_l^u(g, R) < \epsilon$. We can forget P and Q and work solely with R.

Let $I \in R$. For $x, y \in I$, we have
$$\begin{aligned} |(f(x) + g(x)) - (f(y) + g(y))| &= |f(x) - f(y) + g(x) - g(y)| \\ &\leq |f(x) - f(y)| + |g(x) - g(y)| \\ &\leq \operatorname{var}(f, I) + \operatorname{var}(g, I) \end{aligned}$$

It follows that $\text{var}(f+g, I) \leq \text{var}(f, I) + \text{var}(g, I)$. Then

$$\Sigma_l^u(f+g, R) = \sum_{I \in R} \text{var}(f+g, I) \cdot |I| \leq \sum_{I \in R} (\text{var}(f, I) + \text{var}(g, I)) \cdot |I|$$

$$= \sum_{I \in R} \text{var}(f, I) \cdot |I| + \sum_{I \in R} \text{var}(g, I) \cdot |I|$$

$$= \Sigma_l^u(f, R) + \Sigma_l^u(g, R) < \epsilon + \epsilon$$

We see that $\Sigma_l^u(f+g, P)$ can be made arbitrarily small, and so $f+g$ is integrable.

As for $f \cdot g$, we use the same partition R. To estimate the variation of $f \cdot g$ on $I \in R$. Get a number $A \geq |f(x)|$ and $A \geq |g(x)|$, for all $x \in [a,b]$. Let $x, y \in I$, and compute

$$|f(x) \cdot g(x) - f(y) \cdot g(y)| = |(f(x) - f(y)) \cdot g(x) + f(y) \cdot (g(x) - g(y))|$$

$$\leq |f(x) - f(y)| \cdot |g(x)| + |f(y)| \cdot |g(x) - g(y)|$$

$$\leq \text{var}(f, I) \cdot A + A \cdot \text{var}(g, I)$$

It follows that

$$\Sigma_l^u(f \cdot g, R) \leq A \cdot \Sigma_l^u(f, R) + A \cdot \Sigma_l^u(g, R) \leq 2 \cdot A \cdot \epsilon$$

This proves that $f \cdot g$ is integrable. □

Constant functions are integrable, being continuous or increasing – take your pick. Proposition 6.7 then shows that constant multiples of an integrable function are integrable.

Another result at this same level.

*PROPOSITION 6.8. Let $f(x)$ be integrable on $[a, b]$. Then $|f(x)|$ is integrable on $[a, b]$.

PROOF. Given $\epsilon > 0$, let P be a partition of $[a,b]$ such that $\Sigma_l^u(f,P) < \epsilon$. For $I \in P$ and $x, y \in I$, the triangle inequality shows that

$$|f(x)| - |f(y)| \leq |f(x) - f(y)| \leq \text{var}(f, I)$$

and

$$|f(y)| - |f(x)| \leq |f(x) - f(y)| \leq \text{var}(f, I)$$

so that

$$\bigl||f(x)| - |f(y)|\bigr| \leq \text{var}(f, I)$$

It follows that $\text{var}(|f|, I) \leq \text{var}(f, I)$. Thus, $\Sigma_l^u(|f|, P) < \epsilon$. □

Next we show how being integrable on a closed interval is related to being integrable on sub-intervals. You will expect an equation relating the three integrals in the following; that identity will be proved later.

*PROPOSITION 6.9. *Let $a \leq b \leq c$. Then the function $f(x)$ is integrable on $[a, c]$ if and only if it is integrable on $[a, b]$ and integrable on $[b, c]$.*

PROOF. Let $f(x)$ be integrable on $[a, c]$ (the whole interval). Choose $\epsilon > 0$ and let P be a partition of $[a, c]$ such that $\Sigma_l^u(f, P) < \epsilon$. Let R be a partition of $[a, c]$ that refines P and the partition $\{[a, b], [b, c]\}$. In particular, R can be broken into two subsets Q and T such that Q is a partition of $[a, b]$ and T is a partition of $[b, c]$. Observe that

$$\Sigma_l^u(f, R) = \Sigma_l^u(f, Q) + \Sigma_l^u(f, T)$$

and since the three numbers in this equation are non-negative, we also have

(6.4) $\quad \Sigma_l^u(f, Q) \leq \Sigma_l^u(f, R) \quad \text{and} \quad \Sigma_l^u(f, T) \leq \Sigma_l^u(f, R)$

Since R refines P, Proposition 6.6 shows that $\Sigma_l^u(f, R) \leq \Sigma_l^u(f, P)$ so since $\Sigma_l^u(f, P) < \epsilon$ we see that $\Sigma_l^u(f, R) < \epsilon$. Then the inequalities (6.4) lead to $\Sigma_l^u(f, Q) < \epsilon$ and $\Sigma_l^u(f, T) < \epsilon$. That ϵ is arbitrary now proves that $f(x)$ is integrable on $[a, b]$ and on $[b, c]$.

Conversely, if $f(x)$ is integrable on $[a,b]$ and on $[b,c]$, then there are partitions Q of $[a,b]$ and T of $[b,c]$ such that $\Sigma_l^u(f,Q) < \epsilon$ and $\Sigma_l^u(f,T) < \epsilon$ where ϵ is a given positive number. Let $R = Q \cup T$ and then R is a partition of $[a,c]$ for which $\Sigma_l^u(f,R) = \Sigma_l^u(f,Q) + \Sigma_l^u(f,T) < 2 \cdot \epsilon$. □

We close this section by showing that if an integrable function is "messed up" at finitely many points, the resulting function is still integrable. Later, we will discover the surprising fact that the messing up does not change the value of the integral.

*PROPOSITION 6.10. *Let $f(x)$ be integrable on $[a,b]$, and let $g(x)$ be defined on $[a,b]$ with $f(x) = g(x)$ except at finitely many points. Then $g(x)$ is integrable on $[a,b]$.*

PROOF. Let $a_1, a_2, \ldots, a_n$ be the points at which f and g disagree. Put $h(x) = g(x) - f(x)$ so that $h(x)$ is zero except at the a_i. In particular, $h(x)$ is obviously bounded, and it is continuous except at finitely many points. By Proposition 6.3 the function h is integrable. Since $f(x)$ and $h(x)$ are integrable, Proposition 6.7 shows that $g(x) = h(x) + f(x)$ is integrable. □

4. Upper and Lower Sums.

We are ready to define a particular type of Riemann sum, the "upper sum." To do this, we need some notation: for a bounded function f defined on a set I, the symbol $\sup(f, I)$ means the sup of $f(x)$ over $x \in I$. When $f(x)$ is continuous on I, the number $\sup(f, I)$ is the maximum value of f on I.

Let $f : [a,b] \to \mathbb{R}$ be bounded, and let P be a partition of $[a,b]$. Define

$$\Sigma^u(f, P) = \sum_{I \in P} \sup(f, I) \cdot |I|$$

This is called the *upper Riemann sum for f on P*. We will shorten "upper Riemann sum" to "upper sum." As with the variation sum, we will draw a picture suggesting what such a sum looks like.

Now we do the same thing from below rather than from above. If $f(x)$ is bounded on the set I, then $\inf(f, I)$ is the inf of f on I. If $f : [a, b] \to \mathbb{R}$ is bounded and if P is a partition of $[a, b]$, then define

$$\Sigma_l(f, P) = \sum_{I \in P} \inf(f, I) \cdot |I|$$

This is the *lower Riemann sum for f on P*. We will say "lower sum."

Here is how the upper and lower sums behave with respect to refinement.

*PROPOSITION 6.11. *Let $f : [a, b] \to \mathbb{R}$ be bounded. Let P be a partition of $[a, b]$, and let R be a partition of $[a, b]$ refining P. Then*

$$\Sigma_l(f, P) \leq \Sigma_l(f, R) \leq \Sigma^u(f, R) \leq \Sigma^u(f, P)$$

PROOF. We will use the notation of Proposition 6.6 to relate intervals of P to those of R: for each $I \in P$, let I_R be the set of $J \in R$ such that $J \subseteq I$. Then I_R is a partition of I. Each $K \in R$ is in I_R for a unique $I \in P$.

For $I \in P$ and $K \in I_R$ (so that $K \in R$), we have $K \subseteq I$. We claim that $\inf(f, I) \leq \inf(f, K)$. Indeed, let $x \in K$, then $x \in I$ and so $\inf(f, I) \leq f(x)$. We see that $\inf(f, I)$ is a lower bound for the set of $f(x)$ such that $x \in K$, and so $\inf(f, I)$ is less than or equal to the *greatest* lower bound $\inf(f, K)$.

In light of this, we compute

$$\Sigma_l(f,P) = \sum_{I \in P} \inf(f,I) \cdot |I|$$
$$= \sum_{I \in P} \left[\inf(f,I) \cdot \sum_{K \in I_R} |K| \right]$$
$$= \sum_{I \in P} \sum_{K \in I_R} \inf(f,I) \cdot |K|$$
$$\leq \sum_{I \in P} \sum_{K \in I_R} \inf(f,K) \cdot |K|$$
$$= \sum_{K \in R} \inf(f,K) \cdot |K| = \Sigma_l(f,R)$$

so that $\Sigma_l(f,P) \leq \Sigma_l(f,R)$.

The inequality $\Sigma_l(f,R) \leq \Sigma^u(f,R)$ is obvious from the fact that

$$\inf(f,K) \leq \sup(f,K)$$

for each $K \in R$. The proof that $\Sigma^u(f,R) \leq \Sigma^u(f,P)$ depends on the inequality $\sup(f,K) \leq \sup(f,I)$ when $K \subseteq I$. The proof will be left to class or homework. □

A consequence: *every* lower sum is less than or equal to *every* upper sum.

PROPOSITION 6.12. *Let $f : [a,b] \to \mathbb{R}$ be bounded and let P and Q be partitions of $[a,b]$. Then*

$$\Sigma_l(f,P) \leq \Sigma^u(f,Q)$$

PROOF. Let R be a common refinement of P,Q. By Proposition 6.11 we have

$$\Sigma_l(f,P) \leq \Sigma_l(f,R) \leq \Sigma^u(f,R) \leq \Sigma^u(f,Q)$$

as needed. □

5. The Integral.

Here is the crucial link between the value of the integral and the idea of being integrable: the difference between the upper and lower sums for the same partition is the variation sum for that partition.

*PROPOSITION 6.13. *Let $f : [a, b] \to \mathbb{R}$ be bounded, and let P be a partition of $[a, b]$. Then $\Sigma^u(f, P) - \Sigma_l(f, P) = \Sigma_l^u(f, P)$.*

PROOF. For each $I \in P$ Proposition 6.1 proves that $\sup(f, I) - \inf(f, I) = \operatorname{var}(f, I)$. The desired equality follows easily. □

Now we are ready to define the integral of an integrable function. It is the unique number between all the upper and lower sums.

**PROPOSITION 6.14. *Let $f(x)$ be integrable on $[a, b]$. Then there is a unique number I such that $\Sigma_l(f, P) \leq I \leq \Sigma^u(f, P)$ for all partitions P of $[a, b]$.*

PROOF. Let $P = \{[a, b]\}$, the trivial partition of $[a, b]$. Proposition 6.12 shows that $\Sigma_l(f, P) \leq \Sigma^u(f, Q)$ for every partition Q of $[a, b]$. In particular, the upper sums are bounded below by $\Sigma_l(f, P)$; we define I to be the inf of the upper sums. Then $I \leq \Sigma^u(f, Q)$ for all partitions Q.

Now let P be an arbitrary partition of $[a, b]$. Proposition 6.12 shows that $\Sigma_l(f, P) \leq \Sigma^u(f, Q)$ for all partitions Q, so that $\Sigma_l(f, P)$ is a lower bound for all the upper sums. The definition of I as inf shows that $\Sigma_l(f, P) \leq I$, and now we have that I is between all lower and upper sums.

As to uniqueness, if I, J are between all lower and upper sums, then for all partitions P we have

$$|I - J| \leq \Sigma^u(f, P) - \Sigma_l(f, P) = \Sigma_l^u(f, P)$$

using Proposition 6.13. Since $\Sigma_l^u(f,P)$ can be made arbitrarily small, we see that $I = J$. □

As you expect, the number I of Proposition 6.14 is the *definite integral of f over $[a,b]$*. The integral is denoted by the usual notations: $\int_a^b f(x) \cdot dx$ or $\int_a^b f$. The latter notation emphasizes that the integral depends only on the function f and the interval $[a,b]$. Although the differential is customary and useful in manipulations, it is not strictly necessary.

You know that antiderivatives are used to calculate integrals; we will get to that fact soon. It might be useful to compute an integral from the definition right away. Try x^2 on $[0,1]$.

6. Riemann Sums and Integral Algebra.

We are ready to define Riemann sums in general so that we can establish the fundamental formulas for computing integrals.

Suppose that $f : [a,b] \to \mathbb{R}$ and that P is a partition of $[a,b]$. If for each $I \in P$, we choose $x_I \in I$, then a *Riemann sum for f using P* results, having the form

$$\sum_{I \in P} f(x_I) \cdot |I|$$

Here is what is essentially our main theorem: each Riemann sum is within a variation sum of the integral.

**THEOREM 6.15. *Let f be integrable on $[a,b]$, and let P be a partition of $[a,b]$. Let R be a Riemann sum for f using P. Then*

$$\left| R - \int_a^b f \right| \leq \Sigma_l^u(f,P)$$

PROOF. For each $I \in P$, let x_I be chosen from I to form the Riemann sum R. We have $\inf(f,I) \leq f(x_I) \leq \sup(f,I)$, and then

$$\Sigma_l(f,P) \leq R \leq \Sigma^u(f,P)$$

By Proposition 6.14 the integral lies between the lower and upper sum as well. By Proposition 6.13, the lower sum and upper sum are $\Sigma_l^u(f, P)$ apart. Since the integral and the Riemann sum are between two numbers $\Sigma_l^u(f, P)$ apart, we have the desired conclusion. $\square$

Now we will put variation sums, the integral, and Riemann sums together.

*PROPOSITION 6.16. *We have the following, in which it is assumed that the functions mentioned are integrable over the intervals involved.*
a) $\int_a^b (f + g) = \int_a^b f + \int_a^b g$.
b) *If k is a constant, then $\int_a^b (k \cdot f) = k \cdot \int_a^b f$.*
c) *If $a \leq b \leq c$, then $\int_a^c f = \int_a^b f + \int_b^c f$.*
d) *If $f(x) = g(x)$ except at finitely many points, then $\int_a^b f = \int_a^b g$.*

PROOF. We will use the proof of (a) as a model of the argument. Here is an overview: the property we are trying to prove will be obvious for Riemann sums. By Theorem 6.15, the integrals involved will be within a variation sum of the Riemann sums, and so the integrals will "almost" have the property. By the definition of integrable, the variation sums can be made arbitrarily small, and so the integrals will be "arbitrarily close to having the property," and so they will have it! Of course, this paragraph is a *heuristic,* not a proof.

Let's show how it works. For (a), let $\epsilon > 0$. Get a partition A of $[a, b]$ such that $\Sigma_l^u(f, A) < \epsilon$, a partition B such that $\Sigma_l^u(g, B) < \epsilon$, and a partition C such that $\Sigma_l^u(f + g, C) < \epsilon$. If P is a common refinement of A, B, C, then we have $\Sigma_l^u(f, P) < \epsilon$ and $\Sigma_l^u(g, P) < \epsilon$ and $\Sigma_l^u(f + g, P) < \epsilon$.

For each $I \in P$, choose $x_I \in I$, so that the Riemann sum R results for f using P:

$$R = \sum_{I \in P} f(x_I) \cdot |I|$$

Form the Riemann sum R' for g using P by using the same x_I:
$$R' = \sum_{I \in P} g(x_I) \cdot |I|$$
Add these sums:
$$R + R' = \sum_{I \in P} \left[f(x_I) + g(x_I) \right] \cdot |I|$$
and notice that $R + R'$ is a Riemann sum for $f + g$ using P as well. (Here is the reason for using Riemann sums: they are additive, mimicking the property that we want the integral to have.)

We use Theorem 6.15 to estimate
$$\left| \int_a^b (f+g) - \int_a^b f - \int_a^b g \right|$$
$$= \left| \int_a^b (f+g) - (R+R') + R - \int_a^b f + R' - \int_a^b g \right|$$
$$\leq \left| \int_a^b (f+g) - (R+R') \right| + \left| R - \int_a^b f \right| + \left| R' - \int_a^b g \right|$$
$$\leq \Sigma_l^u(f+g, P) + \Sigma_l^u(f, P) + \Sigma_l^u(g, P) < 3\epsilon$$

(This shows that the integrals are "almost additive.") Because ϵ is arbitrary, the original difference in this calculation is 0. (The integrals are "arbitrarily close to being additive," and so they are additive!)

We will be more sketchy about the rest of the conclusions, discussing the details in class. For (b), let $\epsilon > 0$ and get a partition P such that $\Sigma_l^u(f, P) < \epsilon$ and $\Sigma_l^u(k \cdot f, P) < \epsilon$. If we take a Riemann sum for f and multiply by k, we get a Riemann sum for $k \cdot f$, and (b) follows.

We will leave statement (c) to you.

In the setting of (d), get a partition of $[a, b]$ such that $\Sigma_l^u(f, P) < \epsilon$ and $\Sigma_l^u(g, P) < \epsilon$. Because f and g disagree at only finitely many points, we can choose a Riemann sum R for f using P, such that $f(x_I) = g(x_I)$ for all $I \in P$. Then R is a Riemann sum for g as well. $\square$

We call attention to Proposition 6.16d, which says that the integral is not affected by changing finitely many points. This seems to say that the integral computes some sort of "cumulative" or "average" behavior rather than a point by point quantity.

Our next result says that integration preserves inequalities. The inequality 6.17c is called the *triangle inequality*.

*PROPOSITION 6.17. *We have the following.*
a) Let h be integrable and non-negative on $[a,b]$. Then $\int_a^b h \geq 0$.
b) Let f, g be integrable on $[a,b]$ and suppose that $f(x) \leq g(x)$ for all $x \in [a,b]$. Then $\int_a^b f \leq \int_a^b g$.
c) Let f be integrable on $[a,b]$. Then $|\int_a^b f| \leq \int_a^b |f|$.

PROOF. For (a), observe that all lower Riemann sums for h on $[a,b]$ are non-negative.

For (b), the function $g(x) - f(x)$ is non-negative, and so its integral is non-negative by (a). Since

$$0 \leq \int_a^b (g - f) = \int_a^b g - \int_a^b f$$

(by Proposition 6.16a,b), we see that $\int_a^b g \geq \int_a^b f$.

For (c), let $s = \pm 1$ be chosen so that $s \cdot \int_a^b f = |\int_a^b f|$. Then $s \cdot f \leq |f|$, and so by (b) we have

$$\left|\int_a^b f\right| = s \cdot \int_a^b f = \int_a^b s \cdot f \leq \int_a^b |f|$$

□

We have defined the definite integral $\int_a^b$ in the case $a \leq b$. It will be convenient to allow the case $a > b$ as well. We simply define

$$\int_a^b f(x) \cdot dx = -\int_b^a f(x) \cdot dx$$

when $a > b$ and $f(x)$ is continuous on $[b, a]$. You will have an exercise to show that Proposition 6.16abc hold in general.

7. The Fundamental Theorem of Calculus.

The link between antiderivatives and applied quantities (area, volumes, moments, etc.) was observed well before the discovery of the calculus. Both Newton and Leibnitz found ways to elucidate the connection. In this section, we follow a path similar to that of Newton and his predecessors.

If $f(x)$ is integrable on $[a, b]$ and if $t \in [a, b]$, then by Proposition 6.9 the function $f(x)$ is integrable on $[a, t]$. Newton thought up the definition

$$F(t) = \int_a^t f(x) \qquad \text{for} \quad t \in [a, b]$$

We call F the *integral function* of $f(x)$. One way to think of $F(t)$ is as an "area function," since if $f(x) > 0$ on $[a, b]$, then $F(t)$ looks to represent the area under the curve $y = f(x)$ over the interval $[a, t]$.

If c, d are in $[a, b]$, then Proposition 6.16c shows that

$$\int_a^d f(x) = \int_a^c f(x) + \int_c^d f(x)$$

(Recall that this holds if $c \leq d$ and also if $c > d$!) Thus,

(6.5) $$F(d) - F(c) = \int_c^d f(x)$$

From calculus, we are expecting that F is an antiderivative of f. This is not always the case. However, F is always continuous, even if f is not continuous.

**PROPOSITION 6.18. *Assume that $f(x)$ is integrable on $[a, b]$. Let $F(x)$ be the integral function of $f(x)$ on $[a, b]$. Then $F(x)$ is continuous.*

PROOF. The function $|f(x)|$ is bounded above, say by A. Also, Proposition 6.8 says that $|f(x)|$ is integrable. Let $c \leq d$ with $c, d \in [a, b]$, and we combine (6.5) with Proposition 6.17c to estimate

$$|F(d) - F(c)| = \left| \int_c^d f(x) \right| \leq \int_c^d |f(x)|$$

We can estimate the right-most integral using the upper bound A. (There are several ways to justify this estimate; think of one!)

$$\int_c^d |f(x)| \leq A \cdot (d - c)$$

so that $|F(d) - F(c)| \leq A \cdot (d - c) = A \cdot |d - c|$.

The estimate we just obtained for $c \leq d$ can be used for $c \geq d$, and we get $|F(c) - F(d)| \leq A \cdot (c - d) = A \cdot |d - c|$. Thus, we have $|F(d) - F(c)| \leq A \cdot |d - c|$ for all $c, d \in [a, b]$. In other words, F has a Δ-bound! By an exercise, it follows that F is continuous. □

Proposition 6.18 gives rise to an important idea: the integral *smooths a function*; the fact that the integral $F(x)$ is continuous means that F is *smoother* than the possibly discontinuous integrand.

Our next two facts are absolutely essential.

**INTEGRAL MEAN VALUE THEOREM. *Let $f(x)$ be continuous between the numbers a, b. Then there is c between a and b such that*

$$\int_a^b f(x) = f(c) \cdot (b - a)$$

PROOF. Let $a \leq b$. (We will leave the case $a > b$ to class or to you.)

By the Extreme Value Theorem, the function $f(x)$ has a maximum $f(\alpha)$ and a minimum $f(\beta)$ on $[a, b]$. Proposition 6.17b we have

$$f(\beta) \cdot (b - a) \leq \int_a^b f(x) \cdot dx \leq f(\alpha) \cdot (b - a)$$

The function $f(x) \cdot (b-a)$ is continuous, and so the Intermediate Value Theorem finds c between α and β (so that $c \in [a,b]$) such that
$$f(c) \cdot (b-a) = \int_a^b f(x)$$
□

We call attention to the hypothesis that f is continuous in the following.

Fundamental Theorem of Calculus. *Let $f(x)$ be continuous on $[a,b]$. Then the integral $F(x)$ for $f(x)$ on the interval $[a,b]$ is an antiderivative for $f(x)$. If $G(x)$ is an antiderivative of $f(x)$ on $[a,b]$, then*
$$\int_a^b f(x) = G(b) - G(a)$$

PROOF. We begin by showing that $F'(x) = f(x)$. Fix $x \in [a,b]$. For $y \in [a,b]$, the Integral MVT shows that $F(y) - F(x) = f(r) \cdot (y-x)$ for some r between or equal to x, y. Write $r(y)$ for this r to indicate the dependence of y. Also define $r(x) = x$. We claim that $f(r(y))$ is a secant function for $F(x)$ at x. We have only to show that $f(r(y))$ is continuous at x. Let $\epsilon > 0$. Since f is continuous at x, there is $\delta > 0$ such that if $y \in [a,b]$ and $|y - x| < \delta$, then $|f(y) - f(x)| < \epsilon$. For such y, $r(y)$ is also within δ of x, and so
$$|f(r(y)) - f(r(x))| = |f(r(y)) - f(x)| < \epsilon$$
This proves that $f(r(y))$ is continuous at x. Therefore, $F'(x) = f(r(x)) = f(x)$.

If $G(x)$ is an antiderivative of $f(x)$ then the Mean Value Theorem finds a constant C such that $G(x) = F(x) + C$ and then
$$G(b) - G(a) = F(b) + C - (F(a) + C) = F(b) - F(a)$$
$$= F(b) - 0 = \int_a^b f$$
□

The Fundamental Theorem seems to say that

(6.6) $$F(b) - F(a) = \int_a^b f(x) \quad \text{when} \quad F'(x) = f(x)$$

In fact, it *does* say this **if** $f(x)$ is continuous. We want to hedge this fact with two examples. If f is not continuous, but only integrable, we can still define the function $F(y) = \int_a^y f(x)$ so that $F(y)$ is continuous and the integral equation in (6.6) is valid (rather trivially). However, we do not have $F'(x) = f(x)$ except at those x where $f(x)$ is continuous. For example, changing the value of f at some point c does not change the integral (Proposition 6.16d), and so does not change F, but it would invalidate $F'(c) = f(c)$ by messing up the right side of this equation. Thus, the integral of an integrable function does not have to be its antiderivative.

There is a second thing that can go wrong with the formula (6.6). Define

$$F(x) = \begin{cases} x^2 \sin(x^{-2}) & \text{for } x \neq 0 \\ 0 & \text{for } x = 0 \end{cases}$$

It is not too hard to see that $F(x)$ has a derivative for every x. Indeed, $F'(x)$ can be computed from the usual rules when $x \neq 0$. And $F'(0)$ can be computed directly from the definition of the derivative. We get

$$F'(x) = \begin{cases} 2x \sin(x^{-2}) - 2x^{-1} \cos(x^{-2}) & \text{for } x \neq 0 \\ 0 & \text{for } x = 0 \end{cases}$$

The presence of the x^{-1} factor in $F'(x)$ makes it fairly easy to see that $F'(x)$ is not bounded on $[0,1]$, and so it is not integrable there. Thus $\int_0^1 F'(x)$ doesn't make sense, even though F is continuous.

Many of the integration formulas of calculus arise from two computational facts: substitution and parts; these techniques are laid out in exercises.

We do have the following generalization of the Fundamental Theorem.

PROPOSITION 6.19. *Suppose that $f(x)$ has an integrable derivative on $[a,b]$. Then*
$$\int_a^b f'(x) \cdot dx = f(b) - f(a)$$

PROOF. Let $\epsilon > 0$. Let P be a partition with $\Sigma_l^u(f', P) < \epsilon$, and write P's sub-division like this:
$$a = x_0 < x_1 < \cdots < x_{n-1} < x_n = b$$
A Riemann sum for f' using the partition P will be within ϵ of $\int_a^b f'$.

Consider the following calculation that uses a collapsing sum and then the Mean Value Theorem to find $c_j \in [x_{j-1}, x_j]$ for each j with $1 \leq j \leq n$.

$$\begin{aligned}
f(b) - f(a) &= f(x_n) - f(x_0) \\
&= f(x_n) - f(x_{n-1}) + f(x_{n-1}) - f(x_{n-2}) + \cdots \\
&\quad \cdots + f(x_2) - f(x_1) + f(x_1) - f(x_0) \\
&= f'(c_n)(x_n - x_{n-1}) + f'(c_{n-1})(x_{n-1} - x_{n-2}) + \cdots + f'(c_1)(x_1 - x_0)
\end{aligned}$$

We see that $f(b) - f(a)$ is a Riemann sum for $f'(x)$ using P, and so this difference is within ϵ of $\int_a^b f'$. Since ϵ is arbitrary, the difference equals the integral. □

8. Notation and Applications.

The integral notation goes back to Leibnitz, who imagined it representing the area bounded by a closed interval $[a,b]$ on the x-axis, the lines $x = a$ and $x = b$, and the curve $y = f(x)$ above the x-axis. He imagined dividing the area up into *infinitely many, infinitely thin* vertical rectangles, each of width dx (where the differential is idealized as being an "infinitely thin" width). The "height" of an infinitely thin rectangle at the point x would be $f(x)$, and so the total area would be the "infinite sum" of all rectangle-areas $f(x) \cdot dx$. Leibnitz used $\int_a^b$ to denote the sum "over all x in $[a,b]$." The integral sign is a German "s," the first letter in the German word summe (sum). Thus, the notation $\int_a^b f(x) \cdot dx$ was originally meant to denote the sum of "all" products $f(x) \cdot dx$ as x ranges between a and b.

Supposing that $f(x)$ has antiderivative $F(x)$, it is easy to anticipate the Fundamental Theorem. Indeed, writing $\frac{dF}{dx} = f(x)$, the integral becomes

$$\int_a^b f(x) \cdot dx = \int_a^b \frac{dF}{dx} \cdot dx = \int_a^b dF$$

If we think of $\int_a^b dF$ as a sum of small changes in F, then this "sum" ought to be the net change in F across $[a,b]$, which would be $F(b) - F(a)$. Voilà! We are not going to try to make "infinitely thin rectangles" rigorous; we wanted to point out where the differential in the integral came from.

We are in general critical of the way calculus texts present applications of integration. Rather than elucidate our criticisms, we choose some representative examples to show how an application can be approached from an advanced point of view.

Suppose we have a continuous, non-negative function $f(x)$ on $[a,b]$ and we wish to calculate the area A between the closed interval $[a,b]$ on the x-axis, and the curve $y = f(x)$ and the vertical lines $x = a$ and $x = b$. The discussion we are about to give works under at least two sets of axioms: (1) we might assume we understand area intuitively and we are trying to calculate it; (2) we might be thinking about how to define area from scratch.

Let P be a partition of $[a,b]$; for $I \in P$, consider the minimum and maximum of $f(x)$ on I; these numbers are $\inf(f,I)$ and $\sup(f,I)$, respectively. The rectangle of width $|I|$ and height $\inf(f,I)$ fits inside the region $R(I)$ bounded by the interval I on the x-axis and $y = f(x)$ for $x \in I$. If we claim an intuitive understanding of area, we would claim that the area $\inf(f,I) \cdot |I|$ of the rectangle would be less than or equal to the area $A(I)$ of $R(I)$. If we claim to be defining (or discovering) area, we would still want $\inf(f,I) \cdot |I| \leq A(I)$ to be true, since that is a property we would want area to have. Thus, we obtain the inequality $\inf(f, I) \cdot |I| \leq A(I)$ whatever our prior view of area. Similarly, we obtain $A(I) \leq \sup(f, I) \cdot |I|$, since the product on the right gives the area of a rectangle containing $R(I)$. When the areas of all the $R(I)$ are added up, we get the area A of the entire region; again, either because we "know" that area has this property or because we "want to define area to have this property." In any case, we see that the area A is *between* lower and upper sums for the function f. Proposition 6.14 shows that the integral $\int_a^b f$ is the *unique number* between the lower and upper sums. Thus, the area A must be the integral. If we understand A intuitively, then we have just discovered its integral formula; if we are trying to define A, we see that it must be defined as an integral.

For almost all of the applications of integration normally encountered in calculus (or in undergraduate science), a discussion like the one just given for area can be given.

Not all of what are called *applications of integration* need Riemann sums; some simply ask for an antiderivative. For instance, suppose that the x-axis measures inches, and we have a wood log lying on the interval $[1, 3]$ with *density* at point x equal to x^2 pounds per inch. To calculate the total weight of the log, define $W(x)$ to be the weight over the interval $[1, x]$, and *by definition*, the density is the derivative of W. In other words, $W' = x^2$, so that $W = C + x^3/3$ where C is a constant. Since $W(1) = 0$ (Why?), we see that $C = -1/3$, and so $W = (x^3 - 1)/3$. The total weight is $W(3) = 26/3$ pounds.[2]

There are a few applications of integration that require a more subtle approach than either of the two just illustrated. The calculation of arc length is one such application.

---

[2] Many calculus texts would do this problem using Riemann sums to approximate the weight as $x^2 \cdot \Delta x$ over a sub-interval of width Δx in $[1, 3]$. To those texts, we ask, "What is the meaning of density?" If the answer is, as it should be, "It is a derivative," then Riemann sums are not necessary.

We want to calculate the arc length of the curve $y = f(x)$ for $a \le x \le b$. To keep the technical details to a minimum, assume that $f(x)$ has a continuous derivative on $[a, b]$. For a partition P of $[a, b]$, we use the endpoints of the intervals in P to approximate the length of $f(x)$. Write the sub-division for P:

$$a = x_0 < x_1 < \cdots < x_n = b$$

and define

$$L(P) = \sum_{j=1}^{n} \sqrt{(x_j - x_{j-1})^2 + (f(x_j) - f(x_{j-1}))^2}$$

The quantity $L(P)$ is the sum of lengths of line segment along $y = f(x)$ stretching from $(a, f(a))$ to $(b, f(b))$. The definition of *arc length* $L(f, [a, b])$ is that it is the sup of all the $L(P)$. We will prove the familiar formula:

$$L(f, [a, b]) = \int_a^b \sqrt{1 + f'(x)^2} \cdot dx$$

Notice that since $f'(x)$ is continuous, the integrand is continuous.

First note that the triangle inequality in the plane shows that if the partition Q refines the partition P, then $L(P) \le L(Q)$. We will use this presently.

Let P be a partition of $[a, b]$, and let $I \in P$, writing $I = [x_{j-1}, x_j]$. Compute that

$$\sqrt{(x_j - x_{j-1})^2 + (f(x_j) - f(x_{j-1}))^2}$$

can be written like this:

(6.7) $$(x_j - x_{j-1}) \cdot \sqrt{1 + \left[\frac{f(x_j) - f(x_{j-1})}{x_j - x_{j-1}}\right]^2}$$

By the Mean Value Theorem, there is $x_I \in I$ such that

$$f(x_j) - f(x_{j-1}) = f'(x_I) \cdot (x_j - x_{j-1})$$

and when this is substituted into (6.7), remembering that $x_j - x_{j-1} = |I|$, we obtain

(6.8) $$\sqrt{(x_j - x_{j-1})^2 + (f(x_j) - f(x_{j-1}))^2} = |I| \cdot \sqrt{1 + f'(x_I)^2}$$

Applying (6.8) to each $I \in P$, we obtain the following:

$$L(P) = \sum_{I \in P} |I| \cdot \sqrt{1 + f'(x_I)^2}$$

and this is a Riemann sum for $g(x) = \sqrt{1 + f'(x)^2}$ using P. Each Riemann sum is less than or equal to the upper sum for the same partition, and so we have $L(P) \leq \Sigma^u(g, P)$. The partition P refines the partition consisting of the single interval $[a, b]$, and by Proposition 6.11 we have $\Sigma^u(g, P) \leq \sup(g, [a, b]) \cdot (b - a)$. We conclude that $L(P) \leq \sup(g, [a, b]) \cdot (b - a)$. The partition P was arbitrary, and now we see that the $L(P)$ are bounded above; they therefore have a sup, and this proves that the arc length of $f(x)$ on $[a, b]$ exists.

Let $\epsilon > 0$. Because $L = L(f, [a, b])$ is defined as a sup, there is a partition P such that $L - \epsilon < L(P) \leq L$. We can refine P to the partition Q where $\Sigma_l^u(g, Q) < \epsilon$. We have $L(P) \leq L(Q)$, and so we have $L - \epsilon < L(Q) \leq L$. We have shown that $L(Q)$ is a Riemann sum for g and Q, and so Theorem 6.15 proves that

$$\left| L(Q) - \int_a^b \sqrt{1 + f'^2} \right| \leq \Sigma_l^u(g, Q) < \epsilon$$

Thus, we have

$$\left| L - \int_a^b \sqrt{1 + f'^2} \right| = \left| L - L(Q) + L(Q) - \int_a^b \sqrt{1 + f'^2} \right|$$
$$\leq |L - L(Q)| + \left| L(Q) - \int_a^b \sqrt{1 + f'^2} \right|$$
$$< \epsilon + \epsilon$$

That this is true for all ϵ shows that the integral is equal to the arc length L.

Here is another sort of application: the construction of a function. Because of its importance, we will construct the natural logarithm and its inverse, the exponential function.[3]

We start our discussion by forgetting everything we know about e^x and $ln(x)$! The function $1/x$ is continuous on the set of positive reals, and so the Fundamental Theorem gives it an antiderivative $\ln(x)$ there. By subtracting a constant, if necessary, we can assume that $\ln(1) = 0$.

We claim that $\ln(a \cdot b) = \ln(a) + \ln(b)$ for all positive real numbers a, b. Indeed, let $b > 0$ and define $f(x) = \ln(x \cdot b)$ for $x > 0$. Then

$$f'(x) = \frac{1}{x \cdot b} \cdot b = \frac{1}{x}$$

The Mean Value Theorem finds a constant C such that $f(x) = \ln(x) + C$. Plugging in $x = 1$, we see that $\ln(b) = \ln(1) + C = C$, so that $f(x) = \ln(x) + \ln(b)$, and this is the claimed identity. It follows by induction that if n is a positive integer, then $\ln(a^n) = n \cdot \ln(a)$ for all $a > 0$.

Next, we claim that $\ln(x) \to \infty$ as $x \to \infty$. The derivative of $\ln(x)$ shows that it is strictly increasing, so that $\ln(2) > ln(1) = 0$. For each positive integer n, we have $\ln(2^n) = n \cdot \ln(2)$, and so $\ln(2^n) \to \infty$ as $n \to \infty$. Since $\ln(x)$ is increasing, this proves that $\ln(x) \to \infty$ as $x \to \infty$.

For $a > 0$, we have $0 = \ln(1) = \ln(a/a) = \ln(a) + \ln(1/a)$, and we see that $\ln(1/a) = -\ln(a)$. As $a \to 0+$, we have $1/a \to \infty$, and we see that $\ln(1/a) \to \infty$. It follows that $\ln(a) \to -\infty$ as $a \to 0+$.

We have proved that $\ln(x)$ maps the open interval $(0, \infty)$ to the open interval $(-\infty, \infty)$. Since it has a non-zero derivative, the Differentiable IFT shows that it has a differentiable inverse function exp that maps the real numbers one to one, onto the positive real numbers. We have $y = \exp(x)$ if

---

[3]Once it was common to see this construction in elementary calculus. These days, the exponential function is assumed from the start – usually based on an artificial limit definition of the number e and no discussion of the meaning of irrational exponents. Our approach here is much more rigorous.

and only if $x = \ln(y)$, and so $\exp(0) = 1$. We know that $\frac{dy}{dx} \cdot \frac{dx}{dy} = 1$ in this context. It follows that $\exp'(x) = \exp(x)$. The derivative is positive, and so $\exp(x)$ is strictly increasing, and we have the limits

$$\lim_{x \to \infty} \exp(x) = \infty \quad \text{and} \quad \lim_{x \to -\infty} \exp(x) = 0$$

The identity for $\ln(a \cdot b)$ easily leads to $\exp(A + B) = \exp(A) \cdot \exp(B)$ for all $A, B \in \mathbb{R}$. (This latter identity was also an exercise previously.) It follows by induction that if n is a positive integer, then $\exp(n \cdot c) = \exp(c)^n$ for every $c \in \mathbb{R}$ and every positive integer n.

It is an important fact that the exponential and logarithmic functions can be used to define irrational exponents. We have established the identity $a^b = \exp(b \cdot \ln(a))$ when $a > 0$ and b is rational. We *define* a^b to be $\exp(b \cdot \ln(a))$ when b is irrational.

We define $e = \exp(1)$, this number is called *Euler's number*, and our definition of a^b shows that

$$e^x = \exp(x \cdot \ln(e)) = \exp(x) \quad \text{for all} \quad x \in \mathbb{R}$$

Now we have the usual notation for the exponential function.

Many other functions can be built the way we built e^x and $\ln(x)$. We may have time in class to give a couple of examples. Besides being happy about having a secure version of the exponential function, you should notice how the *algebraic* and *graphical* properties of e^x and $\ln(x)$ followed from the *analytic* fact that $1/x$ has an antiderivative.

One final note about applications. You were probably introduced to the connection between Riemann sums and the integral through a limit process that involved the *mesh* – the maximum width of a sub-interval in a partition. One says that the Riemann sums approach the integral as the mesh goes to 0. We have deliberately avoided this approach, but in concession to your past, we will at least state the relevant theorem.

PROPOSITION 6.20. *Let f be integrable on $[a, b]$. For all $\epsilon > 0$, there is $\delta > 0$ such that if P is a partition of $[a, b]$ having mesh less than δ, and if R is a Riemann sum for f using P, then $|R - \int_a^b f| < \epsilon$.*

9. Problems

1. Describe an explicit partition P of $[0, \pi]$ with $\Sigma_l^u(\sin(x), P) < 10^{-6}$. (Hint: use the MVT to estimate the variation of $\sin(x)$ on an interval.)

2. Define $g(x)$ on $[0, 1]$ to be 1 if x is rational and 0 if x is irrational. Show that every variation sum for $g(x)$ is 1. (Thus, $g(x)$ is not integrable.)

3. Let $f(x)$ be bounded on $[a, b]$. Suppose that $f(x)$ is integrable on $[a, c]$ for every c with $a \le c < b$. Show that $f(x)$ is integrable on $[a, b]$. (Hint: choose $a < c < b$ with c "close" to b. Form a partition using a friendly partition of $[a, c]$ along with the interval $[c, b]$.)

4. Complete the steps indicated to compute
$$\int_0^5 x^3 \cdot dx = \frac{625}{4}$$

a) Let P_n be a partition of $[0, 5]$ into n equal width sub-intervals. Compute the upper and lower Riemann sums for this partition. (Hint: the graph shows infs and sups.)

b) Compute the limit as $n \to \infty$ of $\Sigma^u(x^3, P_n)$ and $\Sigma_l(x^3, P_n)$. (Note: you will need a formula for the sum of the first n cubes. Combinatorics text?)

c) Argue from the equality of the limits in (b) that you have the integral.

5. Here is Fermat's idea for computing $\int_0^1 x^k \cdot dx$ where k is a positive integer. Let $0 < r < 1$ and let n be a positive integer. Form the subdivision

$$0 < r^n < r^{n-1} < \cdots < r^2 < r < 1$$

Compute the upper Riemann sum for this subdivision – it's a geometric series, so you should get a "nice" formula! Next let $n \to \infty$. Now let $r \to 1$. What do you get? (Hint: for the last limit $r \to 1$, you might use L'Hôpital.)

6. Prove Proposition 6.16c.

7. Define $f(x)$ on $[0, 1]$ by the formulas:

$$f(x) = \begin{cases} 1/2 & \text{for } 1/2 < x \le 1 \\ 1/4 & \text{for } 1/4 < x \le 1/2 \\ 1/8 & \text{for } 1/8 < x \le 1/4 \\ \vdots & \end{cases}$$

Finally, define $f(0) = 0$. Show that $f(x)$ is integrable on $[0, 1]$ and compute its integral there. (Hint: geometric series.)

8. Show that Proposition 6.16abc hold no matter what is the order relation of the limits of integration. (The functions involved need to be integrable between any limits of integration that occur.)

9. (Substitution.) Let $g : [a, b] \to [c, d]$ and have a continuous derivative. Let $f : [c, d] \to \mathbb{R}$ be continuous. Then

$$\int_a^b f(g(x)) \cdot g'(x) \cdot dx = \int_{g(a)}^{g(b)} f(y) \cdot dy$$

10. (Parts.) Let $f : [a, b] \to \mathbb{R}$ with $f'(x)$ continuous, and let $g : [a, b] \to \mathbb{R}$ be continuous. Let G be an antiderivative of g on $[a, b]$. Then

$$\int_a^b f(x) \cdot g(x) \cdot dx = f(b) \cdot G(b) - f(a) \cdot G(a) - \int_a^b f'(x) \cdot G(x) \cdot dx$$

11. Let $a > 0$ and let b be a rational number. Show that $a^b = \exp(b \cdot \ln(a))$. (Hint: write $b = m/n$ where m, n are integers and $n > 0$; raise a^b and $\exp(b \cdot \ln(a))$ to the n-th power.)

12. Let $a > 0$ and $b, c \in \mathbb{R}$. Show that $a^{b+c} = a^b \cdot a^c$ and that $(a^b)^c = a^{b \cdot c}$. (Hint: use the exponential function, and you can do the irrational and rational cases all at once.)

13. Using the formula $2^x = \exp(x \cdot \ln(2))$, show that
$$\lim_{x \to 0} \frac{2^x - 1}{x} = \ln(2)$$
(In Chapter 5 on p.78, a problem asked you to assume the existence of this limit.)

14. We mentioned in Chapter 5 that $\arctan(x)$ can be defined to have derivative $1/(1+x^2)$. Also, $\arctan(0) = 0$, and so we have
$$\arctan(x) = \int_0^x \frac{dt}{1+t^2} \quad \text{for all} \quad x \geq 0$$
Use this equation to show that $\arctan(x)$ has a numerical limit as $x \to \infty$. (Note: You will prove the *existence* of the limit, not its value – which is $\pi/2$. Do not use the trigonometric functions!) (Hint: integrate from 0 to 1 and then from 1 to x; the second integral can be bounded by estimating the integrand.)

15. Show that $1/\sqrt[3]{1+x^2}$ has an antiderivative $G(x)$ for $x \geq 0$, and with $G(0) = 0$. Explain why $G(x)$ has a differentiable inverse function $H(y)$. Show that $(H')^3 - H^2 = 1$. (Note: this proves that H is what is called an *elliptic function*.)

16. When $x > 0$ and $n \in \mathbb{R}$, write $x^n = \exp(n \cdot \ln(x))$. Use the fact that $\exp' = \exp$ to prove the power rule in this case.

17. Show for each positive integer n that
$$\int_0^\pi \cos^{2k}(\theta) \cdot d\theta = \pi \cdot (-1)^k \cdot \binom{-1/2}{k}$$
(Hint: induction and parts: $\cos(\theta) \cdot \cos^{2k-1}(\theta)$.)

18. Complete the steps to find the arc length of $y = x^2$ for $0 \leq x \leq 1$. Let L be the arc length.

(a) Let $\alpha = \arctan(2)$. Show that $\cos(\alpha) = 1/\sqrt{5}$ and $\sin(\alpha) = 2/\sqrt{5}$.

(b) Use the substitution $\theta = \arctan(2x)$ to change the integral for L in x into one in θ. (The substitution is easier to compute if we write $\tan(\theta) = 2x$.)

(c) Use integration by parts on the integrand: $\sec(\theta) \cdot \sec^2(\theta)$. (We know an antiderivative for $\sec^2(\theta)$.) The quantity L will show up in two places; solve for it.

(d) Show that the derivative of $\ln|\sec(\theta) + \tan(\theta)|$ is $\sec(\theta)$. (This is a standard calculus formula; maybe you remember it.)

(e) Find a formula for L in terms of $\sqrt{5}$ and the logarithm.

CHAPTER 7

Uniform Convergence.

1. The Definition

Nested limits are not always switchable. Compute that

$$\lim_{n \to \infty} \lim_{x \to 1-} x^n = \lim_{n \to \infty} 1 = 1$$
$$\lim_{x \to 1-} \lim_{n \to \infty} x^n = \lim_{x \to 1} 0 = 0$$

Thus, if we have a sequence of functions $f_1(x), f_2(x), \ldots$, and we have the limit

(7.1) $$f(x) = \lim_{n \to \infty} f_n(x) \quad \text{for each} \quad x$$

we cannot necessarily do something like this:

$$\lim_{x \to \alpha} f(x) = \lim_{x \to \alpha} \lim_{n \to \infty} f_n(x) = (??) \lim_{n \to \infty} \lim_{x \to \alpha} f_n(x)$$

In particular, we cannot switch limits with derivatives or integrals – at least not in general.

We indulge in a discussion that should remind you of the beginning of the section on uniform continuity (Chapter 4, Section 4). The limit (7.1) means that for each x and each $\epsilon > 0$, there is N such that if $n \geq N$ then

$$|f_n(x) - f(x)| < \epsilon$$

The N (which is essentially a "delta") is allowed to depend on both x and ϵ. When the dependence on x can be removed, we say that $f_n(x)$ converges *uniformly* to $f(x)$.

Here is the definition made more formally: Let $S \subseteq \mathbb{R}$ and let $f_n : S \to \mathbb{R}$ be a sequence of functions. We say that $f_n(x)$ *converges uniformly to $f(x)$*

on S if for all $\epsilon > 0$ there exists N such that if $n \geq N$ and $x \in S$, then $|f_n(x) - f(x)| < \epsilon$.

In the definition of uniform convergence, the N depends only on ϵ, and not on $x \in S$.

Our first example revisited.

Example. The sequence $f_n(x) = x^n$ does not converge uniformly on $[0, 1]$. The limit function $f(x)$ is 0 on $[0, 1)$ and 1 at $x = 1$. Indeed, let $\epsilon = 1/2$ and suppose that N is given. (We will show that N "doesn't work" for the epsilon.) Let n be a positive integer greater than N. We know that $1/2$ has a positive n-th root α. Since $\alpha^n = 1/2$, we must have $\alpha < 1$. Let x be a number between α and 1, and so $f(x) = 0$. But $x^n > \alpha^n = 1/2$, and so $|f_n(x) - f(x)| > 1/2$.

Here is a famous theorem for recognizing when convergence is uniform. It involves an infinite series of functions. This result, *Weierstrass' Comparison Test* is also called the *M Test*. In the statement of this result, we remind you what is meant by an infinite series.

WEIERSTRASS' COMPARISON TEST. *Let $S \subseteq \mathbb{R}$ and let $f_n : S \to \mathbb{R}$ for each non-negative integer n. Suppose that $|f_n(x)| \leq M_n$ for each n, where M_n is a constant. Suppose also that*

$$\sum_{n=0}^{\infty} M_n = \lim_{k \to \infty} \sum_{n=0}^{k} M_k \quad \textit{converges}.$$

Then

$$\sum_{n=0}^{\infty} f_n(x) \quad \textit{converges uniformly}.$$

PROOF. For each non-negative integer n, define

$$F_n(x) = \sum_{k=0}^{n} f_n(x)$$

1. THE DEFINITION

The conclusion of this result can be stated: F_n converges uniformly on S. We will use Proposition 3.3 to establish this. That proposition shows that the converging sequence

$$P_n = \sum_{k=0}^{n} M_k$$

is a Cauchy sequence. Let $\epsilon > 0$, and get the integer K such that if $n > m > K$, then

$$\epsilon > |P_n - P_m| \geq P_n - P_m = \sum_{k=m+1}^{n} M_k$$

Let $x \in S$. Notice that we have already picked K, and so K does not depend on x. If $n > m > K$, then estimate

$$|F_m(x) - F_n(x)| = \left| \sum_{k=m+1}^{n} f_k(x) \right| \leq \sum_{k=m+1}^{n} |f_n(x)|$$

$$\leq \sum_{k=m+1}^{n} M_n < \epsilon$$

This proves that $F_n(x)$ is a Cauchy sequence for each x. By Proposition 3.3, the sequence $F_n(x)$ then converges, say to $F(x)$.

We claim that $F_n \to F$ uniformly. Indeed, let $\epsilon > 0$ and use the integer K, as above, that does not depend on x. For each $x \in S$, there is an integer $m > K$ such that $|F_m(x) - F(x)| < \epsilon$.

If $n > K$, then[1]

$$|F_n(x) - F(x)| = |F_n(x) - F_m(x) + F_m(x) - F(x)|$$
$$\leq |F_n(x) - F_m(x)| + |F_m(x) - F(x)|$$
$$< 2\epsilon$$

This proves that $F_n \to F$ uniformly. □

---

[1]There is an important subtlety in this argument. The m depends on x, but K does not. The estimate we end up with: $|F_n(x) - F(x)| < 3\epsilon$, does not depend on x, but *obtaining the estimate* depends on possibly different m's for the various x's.

We want to prove two theorems involving uniform convergence. The first says that continuity is preserved.

**Proposition 7.1. *Let $S \subseteq \mathbb{R}$ and let $f_n : S \to \mathbb{R}$ for each non-negative integer n. Suppose that each $f_n(x)$ is continuous. Suppose that $f_n(x) \to f(x)$ uniformly on S. Then $f(x)$ is continuous.*

Proof. Let $\alpha \in S$ and $\epsilon > 0$. Get N so that $n \geq N$ implies that $|f_n(x) - f(x)| < \epsilon$ for all $x \in S$. Choose some such n. The function $f_n(x)$ is continuous at α, and so there is $\delta > 0$ such that if $x \in S$ and $|x - \alpha| < \delta$, then $|f_n(x) - f_n(\alpha)| < \epsilon$. For these values of x, we have

$$|f(x) - f(\alpha)| \leq |f(x) - f_n(x)| + |f_n(x) - f_n(\alpha)| + |f_n(\alpha) - f(\alpha)|$$
$$\leq \epsilon + \epsilon + \epsilon$$

This does it. □

Next we show that integration can be switched with the limit when convergence is uniform, for the conclusion of Proposition 7.2 can be written

$$\lim_{n \to \infty} \int_a^b f_n(x) = \int_a^b \left(\lim_{n \to \infty} f_n(x) \right)$$

*Proposition 7.2. *Suppose, for each $n \geq 0$, that $f_n(x)$ is integrable on $[a, b]$. Suppose that the $f_n(x)$ converge uniformly to $f(x)$ on $[a, b]$. Then $f(x)$ is integrable on $[a, b]$ and*

$$\lim_{n \to \infty} \int_a^b f_n(x) = \int_a^b f(x)$$

Proof. Choose $\epsilon > 0$, and the uniformity of convergence finds N such that if $n \geq N$ and $x \in [a, b]$, then $|f_n(x) - f(x)| < \epsilon$. If $x, y \in [a, b]$, then

$$|f(x) - f(y)| \leq |f(x) - f_n(x)| + |f_n(x) - f_n(y)| + |f_n(y) - f(x)|$$
$$\leq 2\epsilon + |f_n(x) - f_n(y)|$$

If I is a sub-interval of $[a,b]$, it follows that $\text{var}(f, I) \leq \text{var}(f_n, I) + 2\epsilon$.

Since f_n is integrable, there is a partition P of $[a,b]$ such that $\Sigma_l^u(f_n, P) < \epsilon$. Then
$$\Sigma_l^u(f, P) \leq \Sigma_l^u(f_n, P) + 2\epsilon \cdot (b-a) < \epsilon + 2\epsilon \cdot (b-a)$$
Therefore, f is integrable.

Continuing with $n \geq N$, Proposition 6.17 justifies the estimates in the following.
$$\left| \int_a^b f - \int_a^b f_n \right| \leq \int_a^b |f - f_n|$$
$$\leq \int_a^b \epsilon = \epsilon \cdot (b-a)$$
and this establishes the limit of the integrals. $\square$

Some exercises show that the derivative can have trouble even if the convergence is uniform.

2. Problems

1. Prove that the sequence $f_n(x) = x^n$ converges uniformly to $f(x) = 0$ on $[0, 1/2]$.

2. Let $g : [0,1] \to [0,1]$ be continuous. For each positive integer n define $f_n : [0,1] \to [0,1]$ by
$$f_n(x) = \begin{cases} g(x) & \text{if } x > 1/n \\ g(0) & \text{if } x \leq 1/n \end{cases}$$
Show that $f_n(x)$ converges uniformly to $g(x)$ as $n \to \infty$.

3. For each positive integer n, define
$$f_n(x) = \sum_{k=1}^n \frac{\sin(k \cdot x)}{2^k}$$
Show that $f_n(x)$ converges uniformly on $\mathbb{R}$. (Hint: Weierstrass!)

4. For each positive integer n, define
$$f_n(x) = x + \frac{1}{n+1}\cdot(1-x)^{n+1} - \frac{1}{n+1} \qquad \text{when } 0 \leq x \leq 1$$
$$f_n(x) = -\frac{1}{n+1} + \frac{1}{n+1}\cdot(1+x)^{n+1} - x \qquad \text{when } -1 \leq x < 0$$
Prove that each $f_n(x)$ has a continuous derivative on $[-,1,1]$ and that $f_n(x)$ converges uniformly to a *non-differentiable* function $f(x)$ on $[-1, 1]$.

5. Prove that the sequence $f_n(x) = x^n/n$, for $n \geq 1$, of differentiable functions converges uniformly to a differentiable function $f(x)$ on $[0,1]$. But show that
$$\lim_{n\to\infty} f_n'(1) \neq f'(1)$$

6. Let $f_n : [0,1] \to \mathbb{R}$ be integrable for $n \geq 1$, and suppose that $f_n(x)$ converges uniformly to $f(x)$. Define
$$F_n(x) = \int_0^x f_n(t)\cdot dt \quad \text{for} \quad x \in [0,1] \quad \text{and} \quad n \geq 1$$
and show that $F_n(x)$ converges uniformly to $\int_0^x f(t)\cdot dt$. (Note: be careful about the details; Proposition 7.2 is relevant.)

CHAPTER 8

Taylor Series.

1. Functions As Series.

Texts in elementary calculus often convey an erroneous impression concerning the contributions of Newton and Leibnitz to the subject. Indeed, much of the calculus was already known to Fermat, Descartes, and others, years before either Newton or Leibnitz was born. For example, the polynomial differentiation and integration formulas were known to Fermat, who had observed the inverse relation between derivatives and area. It was Newton's teacher, Barrow, who suggested to Newton that he consider this inverse relationship as a starting point for mathematical investigation – this investigation leads to the Fundamental Theorem of Calculus.

If much of the calculus was known, what prevented people from seeing the subject as a whole? One answer is that they did not have a general definition of *function*. The dissimilarity of the polynomials, trigonometric functions, and the exponential and logarithmic functions made each seem a special case arising in its own special context. General differentiation and integration formulas seemed only to apply to polynomials.

In the book *Arithmetica Infinitorum*, published in 1655, Wallis had considered functions which could be written as *power series*.

$$f(x) = \sum_{n=0}^{\infty} a_n x^n$$

Such formulas had been around for a long time; witness the *geometric series* formula known to Archimedes:

$$\frac{1}{1-x} = \sum_{n=0}^{\infty} x^n \qquad \text{for} \qquad -1 < x < 1$$

Wallis assumed that these functions are natural generalizations of polynomials, and so they can be differentiated and integrated in the obvious way. So, for instance,

$$\ln(1-x) = -\int \frac{dx}{1-x} = -\int \sum_{n=0}^{\infty} x^n$$
$$= -\sum_{n=0}^{\infty} \int x^n = -\sum_{n=0}^{\infty} \frac{x^{n+1}}{n+1}$$

And a power series formula for the $\ln(1-x)$ seems to result!

Wallis' ideas led to quite a number of novel formulas, but the informal nature of his proofs bothered some others. His work suggests two important questions: 1) How common are functions that can be represented as power series? 2) Can we actually differentiate and integrate power series as if they were polynomials?

Breakthroughs on the first question came before the second question was adequately understood. For instance, Newton realized that the function $(1+x)^r$ (where r is a rational number) could be represented as a power series. It began to look as if all the common functions could be written as power series, and this turned out to be the case – even the exponential, logarithmic, and trigonometric functions are power series, and so there was a natural way to apply the calculus to these functions as well. It is not too much of an exaggeration to say that calculus seemed to have as its subject, at least initially, those functions that could be written as power series.

The previous discussion is not a thorough history; the point is that power series were central to the discovery and use of the calculus. We will give an

1. FUNCTIONS AS SERIES.

introduction to the serious study of such series, as usual making sure that we prove results of classical importance. The emphasis will be on obtaining function formulas rather than on the convergence of constant term series.

All sequences considered will have domain equal to the set of non-negative integers. Given a sequence a_n, you recall studying the *series for* a_n

$$\sum_{n=0}^{\infty} a_n = \lim_{N \to \infty} \sum_{n=0}^{N} a_n$$

where the limit on the right is the *definition* of the series on the left. Informally, we think of the series as an infinite sum:

$$\sum_{n=0}^{\infty} a_n = a_0 + a_1 + a_2 + a_3 \cdots$$

Remember that this is informal, although it is also helpful.

We need two basic facts about series (discussed but perhaps not proved in calculus). Given a sequence a_n, we define its *absolute partial sums* s_k for each non-negative integer k by

$$s_k = \sum_{n=0}^{k} |a_n|$$

Proposition 8.1a is called the *Divergence Test*.

*PROPOSITION 8.1. *Let* a_n *be a sequence.*
a) *If the series for* a_n *converges, then* $a_n \to 0$ *as* $n \to \infty$.
b) *If the absolute partial sums for* a_n *are bounded, then the series for* a_n *converges. Furthermore, for each* $k \in \mathbb{N}$, *we have*

$$\left| \sum_{n=k}^{\infty} a_n \right| \leq \sum_{n=k}^{\infty} |a_n|$$

PROOF. For (a), assume that the series for a_n converges, and notice that

$$a_{k+1} = \sum_{n=0}^{k+1} a_n - \sum_{n=0}^{k} a_n$$

By the convergence of the series for a_n, the sums on the right side both go to the value of the series as $k \to \infty$; therefore, the term on the left goes to 0. This proves (a).

Let s_k be the sequence of absolute partial sums for a_n, and observe that $s_{k+1} = s_k + |a_{k+1}|$, and so the sequence s_k is an increasing sequence. In (b), this sequence is bounded, and so the Monotone Convergence Theorem shows that it converges to its supremum.

Now consider that $||a_n| - a_n| \leq 2|a_n|$, and so the series

$$\sum_{n=0}^{\infty} (|a_n| - a_n)$$

is bounded as well. This series has non-negative terms, and so the reasoning applied to the s_k works again: the partial sums form an increasing sequence, which, since it is bounded above, converges. Then

$$\sum_{n=0}^{k} a_n = \sum_{n=0}^{k} |a_n| - \sum_{n=0}^{k} (|a_n| - a_n)$$

and since, as $k \to \infty$, the two sums on the right each converge to a limit, the sum on the left converges to a limit as well. This proves (b).

Continuing this setting, the triangle inequality is that

$$\left| \sum_{n=k}^{m} a_n \right| \leq \sum_{n=k}^{m} |a_n|$$

Letting $m \to \infty$, we obtain the inequality in (b). $\square$

Now we are ready to consider the function formed from a sequence a_n and a real number c (the *center*), by the formula

(8.1) $$f(x) = \sum_{n=0}^{\infty} a_n (x - c)^n$$

Repeating the definition of series:
$$\sum_{n=0}^{\infty} a_n(x-c)^n = \lim_{N\to\infty}\left(\sum_{n=0}^{N} a_n(x-c)^n\right)$$
Such a function is called a *Taylor series* about c. Taylor studied such series in the early 1700's, and we will discuss his ideas later. If we let $y = x - c$ in (8.1), we get
$$f(y+c) = \sum_{n=0}^{\infty} a_n y^n$$
which is a Taylor series about 0 (sometimes called a *MacLauren series*). In discussing properties of the general Taylor series, the c is usually irrelevant, and so we will do most of our work without it, considering the series

(8.2) $$f(x) = \sum_{n=0}^{\infty} a_n \cdot x^n$$

2. Radius of Convergence.

We will see that the sequence a_n has a number associated with it: the *radius of convergence* R, such that the series (8.2) converges if $|x| < R$ and diverges if $|x| > R$. You probably recall using the ratio test to get the R. There are sequences for which the ratio test does not yield a limit, and so we will need a more general approach. For a given sequence a_n, we need to consider the set of non-negative real numbers r such that $|a_n|r^n$ is bounded above. This set is not empty, since 0 is an element (i.e. $|a_n|0^n$ is bounded above!). Here are the facts we need; some of them will be proved in class and some will show up as homework.

PROPOSITION 8.2. *Let a_n and b_n be sequences.*

a) *Define $a_n = a \cdot s^n$ where a, s are non-zero real numbers. Let $r \geq 0$. Then $|a_n|r^n$ is bounded above if and only if $0 \leq r \leq 1/|s|$.*

***b)** If $0 \leq s \leq r$ and $|a_n|r^n$ is bounded above, then $|a_n|s^n$ is bounded above.*

c) *If $r \geq 0$ and $|a_n|r^n$ is bounded above and $|b_n|r^n$ is bounded above, then $|a_n + b_n|r^n$ is bounded above.*

d) *If $r \geq 0$ and $s \geq 0$ and $|a_n|r^n$ is bounded above and $|b_n|s^n$ is bounded above, then $|a_n b_n|(rs)^n$ is bounded above.*

Now we define the number that controls convergence of Taylor series. For a sequence a_n, let S be the set of all $r \geq 0$ such that $|a_n|r^n$ is bounded above. If S is bounded above, the sup of S is called the *radius of convergence* of a_n. If S is not bounded above, we say that ∞ is the radius of convergence of a_n. We will encounter the phrase: let R be the radius of convergence of a_n and let $|x| < R$. In the case that $R = \infty$, the inequality $|x| < R$ means only that x is a real number.

For a sequence a_n we have defined the set of r such that $a_n \cdot r^n$ is bounded, and we have defined the radius of convergence R. We show that the set of r-values is either $[0, R]$ or $[0, R)$.

****PROPOSITION 8.3.** *Let a_n be a sequence with radius of convergence R. If $r \geq 0$ and $|a_n|r^n$ is bounded above, then $0 \leq r \leq R$. If $0 \leq r < R$, then $|a_n|r^n$ is bounded above.*

PROOF. The first statement is obvious from the definition of R. Let $0 \leq r < R$. If $R = \infty$, then the set S, defined above, is not bounded above and there is $s > r$ such that $|a_n|s^n$ is bounded above. On the other hand, if R is a number, then since it is the sup of S, there is $s > r$ where $|a_n|r^n$ is bounded above. In any case we have $r < s$ where $|a_n|s^n$ is bounded above. Now $|a_n|r^n$ is bounded above by Proposition 8.2b. □

2. RADIUS OF CONVERGENCE.

The most common way to determine the radius of convergence for a sequence is to use the "Ratio Test." We caution you that the way we will state this test involves a ratio that is upside down from the one that occurs in calculus texts. Our ratio more readily gives the r's such that $|a_n|r^n$ is bounded above.

*PROPOSITION 8.4. *Let a_n be a sequence which is eventually non-zero, and suppose that*

$$(8.3) \qquad \lim_{n \to \infty} \left| \frac{a_n}{a_{n+1}} \right| = R$$

Then R is the radius of convergence of a_n (even in the case that the limit diverges to infinity).

PROOF. Let r be a non-negative real number with $r < R$. We will show that $|a_n|r^n$ is bounded above. The limit (8.3) shows that there is a non-negative integer N such that $n \geq N$ implies that

$$\left| \frac{a_n}{a_{n+1}} \right| > r$$

(Observe that this makes sense even when $R = \infty$.) It is an exercise to show that the sequence $|a_n| \cdot r^n$ is bounded above for all $n \geq N$. Since there are only finitely many $|a_n| \cdot r^n$ for $n < N$, we see that $|a_n| \cdot r^n$ is bounded, as needed.

By Proposition 8.3 we see that R is less than or equal to the radius S of convergence of a_n. Assume that $R < S$, and let $R < r < S$. Proposition 8.3 shows that $|a_n| \cdot r^n$ is bounded above. The ratio limit of the a_n finds a non-negative integer N such that $n \geq N$ implies that

$$\left| \frac{a_n}{a_{n+1}} \right| < r \qquad \text{or} \qquad |a_n| < r|a_{n+1}|$$

We can also write this as $|a_n|r^n < r^{n+1}|a_{n+1}|$. In other words, the sequence $|a_n|r^n$ is eventually increasing. It is also bounded above, and therefore it converges to a (positive) limit L. Then

$$R = \lim_{n \to \infty} \left| \frac{a_n}{a_{n+1}} \right| = \lim_{n \to \infty} \frac{r \cdot |a_n| r^n}{|a_{n+1}| r^{n+1}} = \frac{rL}{L} = r$$

This is a contradiction. We have that $R = S$. □

Consider the two sequences $a_n = 1$ and $b_n = n$; in each case the ratio limit is 1, and so the radius of convergence is 1. The sequence $|a_n| 1^n = 1$ is bounded above; the sequence $|b_n| 1^n = n$ is not bounded above. Thus, we may or may not have boundedness at the radius of convergence.

Here is a corollary (really two corollaries) of Propositions 8.4 and 8.3.

COROLLARY 8.5. *If k is a positive integer, then the radius of convergence of the sequence n^k is 1. Let a_n be a sequence with radius of convergence R. Then the sequence $n \cdot a_n$ has radius of convergence R.*

PROOF. The first statement is proved from Proposition 8.4 by the limit

$$\lim_{n \to \infty} \frac{n^k}{(n+1)^k} = 1$$

Let R be the radius of convergence for na_n, and S the radius for a_n. We claim that $R \leq S$. If not, then $S < R$, and there is r with $S < r < R$. By Proposition 8.3, we have that $|na_n| r^n$ is bounded above. But then so is the sequence $|a_n| r^n$. By Proposition 8.3, we have that $r \leq S$, and this is a contradiction.

We have that $R \leq S$. Assume that $R < S$. We will produce positive numbers r and s with $R < rs$ and such that the sequence nr^n is bounded above, the sequence $|a_n| s^n$ is bounded above. It will follow from Proposition 8.2c that $|na_n|(rs)^n$ is bounded above. That $R < rs$ will then be a contradiction, and we will know that $R = S$.

We need r and s. Since $R < S$, there is a number s with $R < s < S$. Since R/s is less than 1, there is a number r with $R/s < r < 1$. We have what we wanted. $\square$

3. Convergence, Continuity, Differentiation.

The radius of convergence was introduced in order to prove the following.

*THEOREM 8.6. *Let a_n be a sequence with radius of convergence R. Then the series $f(x) = \sum_{n=0}^{\infty} a_n x^n$ converges when $|x| < R$ and it diverges when $|x| > R$.*

But we need more. We want to be able to take the derivative and antiderivative of a series, and to do that, we will need the following result, which includes Theorem 8.6. This is one of the most important theorems in the course.

**THEOREM 8.7. *Let a_n be a sequence with radius of convergence R. Let r be a positive real number with $r < R$. Then the series $f(x) = \sum_{n=0}^{\infty} a_n x^n$ converges uniformly on $[-r, r]$. In particular, $f(x)$ converges when $|x| < R$. The series $f(x)$ diverges when $|x| > R$.*

PROOF. Because $r < R$, there is a number s with $r < s < R$. By Proposition 8.3, $|a_n| s^n$ is bounded above; say $|a_n| s^n \leq C$ for all n. Choose x with $|x| \leq r$, and for a non-negative integer n, we estimate

$$|a_n x^n| = |a_n||x^n| = |a_n| s^n \cdot \frac{r^n}{s^n} \cdot \frac{|x|^n}{r^n} \leq C \cdot \left(\frac{r}{s}\right)^n \cdot 1$$

The positive ratio r/s is less than 1, and so the geometric series for $(r/s)^n$ converges. By Weierstrass' Comparison Test, the series for $a_n \cdot x^n$ converges uniformly. In other words, $f(x)$ converges uniformly on $[-r, r]$.

If $|x| < R$, then there there is $r < R$ with $|x| < r$. By the previous argument, f converges on $[-r, r]$; in particular, it converges at x.

We have left to consider the case that $|x| > R$. The sequence $a_n x^n$ is not bounded in this case, and so by Proposition 8.1 the series $f(x)$ diverges. □

We call your attention to another technicality: the hypothesis of Theorem 8.7 cannot hold when $R = 0$, since there cannot be a positive $r < R$. Also, notice that if $R = \infty$, then every positive real number r satisfies the hypothesis of Theorem 8.7. Does this say that the series converges uniformly on the entire real numbers? No! Only on every closed interval on the reals. The difference is subtle and we might have time to say something about it in class.

The uniformity of Theorem 8.7 allows us to permute an integral sign and the series sign. In other words, we can integrate a convergent series term by term. We will show that we can also differentiate term by term, but recall that this does not follow directly from uniform convergence, and so we will need an indirect argument for differentiability.

*THEOREM 8.8. *Let a_n be a sequence with radius of convergence R, and define $f(x) = \sum_{n=0}^{\infty} a_n x^n$. Then $f(x)$ is differentiable for $|x| < R$. In fact,*

$$f'(x) = \sum_{n=1}^{\infty} n \cdot a_n \cdot x^{n-1} \quad \text{and} \quad \left(\sum_{n=0}^{\infty} a_n \cdot \frac{x^{n+1}}{n+1} \right)' = f(x) \quad \text{when} \quad |x| < R$$

PROOF. If $|t| < R$, then there is a real number r with $|t| \leq r < R$, and the series $f(x)$ converges uniformly on $[-r, r]$ by Theorem 8.7. The partial sums of the series are polynomials, which are continuous, and so since the series converges uniformly, Proposition 7.1 shows that it converges to a continuous function. In particular, the series is continuous at t. Furthermore, Proposition 7.2 allows us to permute the limit and integral in the following:[1]

---

[1] Recall that we have defined what $\int_0^t$ means even when $t < 0$.

$$\int_0^t \left(\sum_{n=0}^{\infty} a_n \cdot x^n\right) dx = \int_0^t \left(\lim_{N \to \infty} \sum_{n=0}^{N} a_n \cdot x^n\right) dx$$

$$= \lim_{N \to \infty} \left(\int_0^t \left[\sum_{n=0}^{N} a_n \cdot x^n\right] dx\right)$$

$$= \lim_{N \to \infty} \left(\sum_{n=0}^{N} a_n \cdot \frac{t^{n+1}}{n+1}\right)$$

$$= \sum_{n=0}^{\infty} a_n \cdot \frac{t^{n+1}}{n+1}$$

The Fundamental Theorem of Calculus shows that the resultant series is an antiderivative of f (using the variable t in place of x).

Now consider the series

$$\sum_{n=1}^{\infty} n \cdot a_n \cdot x^{n-1}$$

Corollary 8.5 shows that its radius of convergence is R, and so by Theorem 8.7 it converges on the open interval $(-R, R)$. If x is in this interval, then by the first part of the proof, we can compute

$$\int_0^x \sum_{n=1}^{\infty} n \cdot a_n \cdot t^{n-1} \cdot dt = \sum_{n=0}^{\infty} a_n \cdot x^n$$

By the Fundamental Theorem of Calculus, the integrand on the left is the derivative of $f(x)$ of the right. □

An incredible number of function formulas pour out of Theorem 8.8. In a later section, we will give several examples; for now we continue with technical facts.

If we have the formula (8.1) for $|x - c| < R$ where R is the radius of convergence of the a_n, then, as we will now see, Theorem 8.8 shows how to find the a_n in terms of the derivatives of $f(x)$. We successively calculate, $f(x)$, $f'(x)$, $f''(x)$, and in each case we plug in $x = c$.

$$f(x) = a_0 + a_1 \cdot (x-c) + a_2 \cdot (x-c)^2 + a_3 \cdot (x-c)^3$$
$$+ a_4 \cdot (x-c)^4 + a_5 \cdot (x-c)^5 + \cdots$$

so that $f(c) = a_0$

$$f'(x) = a_1 + 2a_2 \cdot (x-c) + 3a_3 \cdot (x-c)^2$$
$$+ 4a_4 \cdot (x-c)^3 + 5a_5 \cdot (x-c)^4 + \cdots$$

so that $f'(c) = a_1$

$$f''(x) = 2a_2 + 6a_3 \cdot (x-c) + 12a_4 \cdot (x-c)^2 + 20a_5 \cdot (x-c)^3 + \cdots$$

so that $f''(c) = 2a_2$

$$f^{(3)}(x) = 6a_3 + 24a_4 \cdot (x-c) + 60a_5 \cdot (x-c)^2 + \cdots$$

so that $f^{(3)}(c) = 6a_3$

$$f^{(4)}(x) = 24a_4 + 120a_5 \cdot (x-c) + \cdots$$

so that $f^{(4)}(c) = 24a_4$

The general pattern is that $f^{(n)}(c) = n! \cdot a_n$, and so we arrive at the following formula, sometimes called *Taylor's formula*

$$a_n = \frac{f^{(n)}(c)}{n!} \quad \text{for} \quad n \geq 0$$

We emphasize a point of logic: Taylor's formula **assumes** that the function $f(x)$ is represented as in (8.1). If we are searching for a Taylor series for a function, then Taylor's formula tells us what the coefficients have to be, but it **does not** prove that the Taylor series converges or that it converges to the function. For example, if $f(x) = e^{-1/x^2}$ (and set $f(0) = 0$), and if $c = 0$ is used, then it is not too hard to show that the a_n are all 0. The series (8.1) converges, but it does not equal $f(x)$ when $x \neq 0$. If you take $f(x) = 1/(1-x)$ and $c = 0$ and $x = 5$, the series does not converge (it is the geometric series, which only converges when $|x| < 1$). Thus, the formula for a_n does not have to work. It is interesting, however, that these a_n result in an approximation formula for $f(x)$. This is sometimes called *Taylor's Theorem with Remainder*. Notice that the number d in the statement depends on c and x.

*Theorem 8.9. *Let N be a non-negative integer and suppose that $f(x)$ has $N+1$ derivatives in an open interval I. Let $c \in I$. Then for each $x \in I$, there exists d between c and x such that*

$$f(x) = \sum_{n=0}^{N} \frac{f^{(n)}(c)}{n!}(x-c)^n + \frac{f^{(N+1)}(d)}{(N+1)!}(x-c)^{N+1}$$

Proof. Fix $x \in I$. If $x = c$, then the desired equation can be obtained with $d = c$. Suppose that $x \neq c$ and consider the equation in unknown M:

$$f(x) = \sum_{n=0}^{N} \frac{f^{(n)}(c)}{n!}(x-c)^n + \frac{M}{(N+1)!}(x-c)^{N+1}$$

Because $x \neq c$, this equation can be solved for M, and so there is a unique value of M making the equation true.

Now, for $t \in I$, set

$$g(t) = \sum_{n=0}^{N} \frac{f^{(n)}(t)}{n!}(x-t)^n + \frac{M}{(N+1)!}(x-t)^{N+1}$$

Observe that $g(x) = f(x)$ and that $g(c) = f(x)$, the latter equality from the definition of M. The fact that f has $N+1$ derivatives in I shows that g is differentiable, and a calculation using the product rule gives:

$$g'(t) = f'(t) + \sum_{n=1}^{N} \left(\frac{f^{(n+1)}(t)}{n!}(x-t)^n - \frac{f^{(n)}(c)}{(n-1)!}(x-t)^{n-1} \right) - \frac{M}{N!}(x-t)^N$$

One sees that this sum collapses to

$$g'(t) = \frac{f^{(N+1)}(t)}{N!}(x-t)^N - \frac{M}{N!}(x-t)^N$$

We already had that $g(x) = f(x) = g(c)$, and so the Mean Value Theorem gives d between x and c such that $g'(d) = 0$. Because $d \neq x$, the equation for $g'(t)$ shows that $f^{(N+1)}(d) = M$, as needed. $\square$

Suppose we are given $f(x)$ with infinitely many derivatives in the open interval I, with $c \in I$. We see that $f(x)$ can be represented as in (8.1) on p.128 if and only if
$$\lim_{N \to \infty} \frac{f^{(N+1)}(d)}{(N+1)!}(x-c)^{N+1} = 0$$
Since we don't know what d is, this is usually a little slippery. If there is an estimate on the $(N+1)$-st derivative of f, we might be ok.

The error term in Theorem 8.9, the term appearing in the limit of the last paragraph, is called the *remainder*. There are several representations of the remainder, that of Theorem 8.9 is due to Lagrange. Here is a form due to Cauchy:
$$\int_c^x \frac{f^{(N+1)}(t)}{(N+1)!} \cdot (x-t)^N \cdot dt$$
This form of the remainder may be established by integration by parts; we will not give the proof here.

4. Abel's Theorem on the Endpoints.

The series in Theorem 8.7 converges inside its radius of convergence R and diverges outside that radius. This leaves open the question what happens at the *endpoints* $\pm R$. As we will observe from examples, a series can converge or diverge at either or both of these endpoints. In many uses of series, it is not important what happens at the endpoints, but there is an interesting theorem of Abel that says that a series which converges at one of its endpoints is continuous there. This very clever proof involves "summation by parts" analogous to integration by parts.

4. ABEL'S THEOREM ON THE ENDPOINTS.

*THEOREM 8.10. *Let a_n be a sequence with radius of convergence R. Let $r \in \{R, -R\}$, and let I be the closed interval having 0 and r as endpoints. If $f(x) = \sum_{n=0}^{\infty} a_n x^n$ converges at $x = r$, then the convergence is uniform on I, so that $f(x)$ is continuous on I.*

PROOF. If $r = 0$, then the domain of $f(x)$ is $\{0\}$, and so f is continuous at r trivially.

Assume that $r \neq 0$. We will show that $f(x)$ is uniformly convergent on I, and Proposition 7.1 will then show that $f(x)$ is continuous on I.

The fact that the series for $f(x)$ converges at r means that the sequence

$$b_n = \sum_{k=0}^{\infty} a_k \cdot r^k - \sum_{k=0}^{n-1} a_k \cdot r^k = \sum_{k=n}^{\infty} a_k \cdot r^k$$

converges to 0. Thus, if $\epsilon > 0$ is given, there is N such that if $n \geq N$, then $|b_n| < \epsilon$. Keep this in mind. Also, notice that $a_k \cdot r^k = b_k - b_{k+1}$.

Let $x \in I$, and define $y = x/r$. Notice that $y \in [0,1]$ whether $r > 0$ or $r < 0$. Because the series for $f(x)$ converges, the series $\sum_{k=n}^{\infty} a_k \cdot x^k$ converges (as in the definition of b_n). We manipulate a partial sum of this series – here is where summation by parts occurs. Let $m \geq n$.

$$\sum_{k=n}^{m} a_k \cdot x^k = \sum_{k=n}^{m} a_k \cdot r^k \cdot y^k = \sum_{k=n}^{m} (b_k - b_{k+1}) \cdot y^k$$

$$= b_n \cdot y^n + \sum_{k=n+1}^{m} b_k \cdot (y^k - y^{k-1}) - b_{m+1} \cdot y^m$$

Recalling the properties of b_n, we can begin with the foregoing equation along with (8.6) to estimate

$$\left|\sum_{k=n}^{m} a_k \cdot x^k\right| = \left|b_n \cdot y^n + \sum_{k=n+1}^{m} b_k \cdot (y^k - y^{k-1}) - b_{m+1} \cdot y^m\right|$$

$$\leq |b_n \cdot y^n| + \sum_{k=n+1}^{\infty} |b_k \cdot (y^k - y^{k-1})| + |b_{m+1} \cdot y^m|$$

$$\leq |b_n| + \sum_{k=n+1}^{m} |b_k| \cdot |y^k - y^{k-1}| + |b_{m+1}|$$

$$\leq \epsilon + \epsilon \cdot \sum_{k=n+1}^{m} |y^k - y^{k-1}| + \epsilon \leq 3\epsilon$$

This estimate works for all $m \geq n$, and so we can let $m \to \infty$ to conclude that

$$\left|\sum_{k=n}^{\infty} a_k \cdot x^k\right| \leq 3\epsilon$$

This estimate depends only on $n \geq N$ (chosen from the series for $f(r)$) and for all $x \in I$. Since

$$\sum_{k=n}^{\infty} a_k \cdot x^k = \sum_{k=0}^{\infty} a_k \cdot x^k - \sum_{k=0}^{n-1} a_k \cdot x^k$$

we see that the convergence of the series to $f(x)$ is uniform on I, as claimed. □

5. Favorite Examples.

We have already constructed the exponential function. Here, we re-do the construction using series rather than integration, and we will work speculatively – as if we are trying to find a function $\exp(x)$ such that $\exp'(x) = \exp(x)$ and $\exp(0) = 1$.

We will try to find $\exp(x)$ as a series:

$$\exp(x) = \sum_{n=0}^{\infty} a_n \cdot x^n$$

5. FAVORITE EXAMPLES.

In other words, we're trying to *solve* for the coefficients a_n. The condition $\exp(0) = 1$ shows that $a_0 = 1$. Imagining that we are inside an as-yet-to-be-determined radius of convergence, the formula $\exp'(x) = \exp(x)$ leads to

$$a_{n+1} = \frac{a_n}{n+1} \quad \text{for} \quad n = 0, 1, 2, \ldots$$

It follows easily that $a_n = 1/n!$ for all n. Remember that we are working speculatively, but now we have enough information to use the series as the *definition* of $\exp(x)$ and work forward. So, define

$$\exp(x) = \sum_{n=0}^{\infty} \frac{x^n}{n!}$$

The ratio test shows that the series has infinite radius of convergence, and so our definition works for all $x \in \mathbb{R}$. Furthermore, Proposition 8.8 shows that we can differentiate term by term, and this shows that $\exp'(x) = \exp(x)$. A problem in Chapter 5 showed that the algebraic properties of the exponential *follow* from the analytic property $\exp' = \exp$. It also follows that $\exp(x) > 0$ for all x, and so the Differentiable IFT produces a differentiable inverse function: this gives us the function $\ln(x)$. This gets us the same functions and properties as were derived in Chapter 6, where the other construction was done.

Another example of a series arising out of a differential equation is afforded by the binomial series – a historically important example, since Newton's discovery of it helped to suggest the universal nature of the calculus. We seek a series centered about 0 for $(1+x)^\alpha$ where α is an arbitrary real number. In Chapter 5 we remarked that A^α can be defined for an arbitrary positive number A and arbitrary real number α as $\exp(\alpha \cdot \ln(A))$, and that when α is rational, this formula agrees with the algebraic definition of the exponent. We have also seen that the power rule works, so if we let $A = 1 + x$, we see that

$y = (1+x)^\alpha$ satisfies the differential equation

(8.4) $\qquad (1+x) \cdot f'(x) = \alpha \cdot f(x) \qquad$ and $\qquad f(0) = 1$

In the proof of the Binomial Theorem we will build a series that satisfies this same equation, and it will follow that the series represents $(1+x)^\alpha$.

The formula in the following theorem is the familiar algebra formula when α is a non-negative integer. In that case, the formula holds for all real numbers x (the series is a finite sum and has radius infinity).

*BINOMIAL THEOREM. Let $\alpha \in \mathbb{R}$. If α is a non-negative integer, let $I = \mathbb{R}$; otherwise, let I be the open interval $(-1, 1)$. Then we have

$$(1+x)^\alpha = \sum_{n=0}^{\infty} \binom{\alpha}{n} x^n \quad \text{for} \quad x \in I$$

PROOF. Define

$$f(x) = \sum_{n=0}^{\infty} \binom{\alpha}{n} \cdot x^n$$

You have shown that the radius of convergence of the binomial sequence $\binom{\alpha}{n}$ is infinity if α is a non-negative integer, and it is 1 otherwise. Thus, Theorem 8.7 shows that $f(x)$ is defined on the set I defined from α in the hypothesis, and Theorem 8.8 shows that we can differentiate the series term by term on I. In class we will derive from this the fact that $f(x)$ satisifies (8.4). (You will need to recall the recursive equations for $\binom{\alpha}{n}$; these equations were introduced in Chapter 2.)

The differential equation satisfied by $f(x)$ shows that

$$\left(f(x) \cdot (1+x)^{-\alpha} \right)' = 0$$

so that this product is a constant. Since $f(0) = 1$, we easily derive that $f(x) = (1+x)^\alpha$, as claimed. $\qquad \square$

5. FAVORITE EXAMPLES.

The trigonometric functions can also be discovered as series from their differential equations. We will take this approach in class. Here, we will illustrate an alternative approach; we regard $\sin(x)$ as a given function and try to find a Taylor series for it. Theorem 8.9 shows how to approximate $\sin(x)$ by a polynomial of arbitrary degree; we might try to let the degree go to infinity. Using 0 as center, it is easy to use Taylor's formula to calculate

$$\frac{\sin^{(n)}(0)}{n!} = \begin{cases} 0 & \text{if } n \text{ is even} \\ \frac{1}{n!} & \text{if } n-1 \text{ is divisible by 4} \\ -\frac{1}{n!} & \text{if } n-1 \text{ is even but not divisible by 4} \end{cases}$$

It is customary to represent these coefficients using $2k+1$, for $k \geq 0$, to represent the odd numbers. Theorem 8.9 looks like this: there is a number d between x and 0 such that

$$(8.5) \qquad \sin(x) = \sum_{k=0}^{m} \frac{(-1)^k}{(2k+1)!} \cdot x^{2k+1} + \frac{\sin^{(2m+2)}(d)}{(2m+2)!} x^{2m+2}$$

The $(2m+2)$ derivative of $\sin(x)$ is $\pm \cos(x)$ which is bounded above in absolute value by 1. Furthermore, we know that

$$\lim_{m \to \infty} \frac{x^{2m+2}}{(2m+2)!} = 0$$

And so we see that

$$\sin(x) = \sum_{k=0}^{\infty} \frac{(-1)^k}{(2k+1)!} \cdot x^{2k+1}$$

for all $x \in \mathbb{R}$. The series for $\cos(x)$ can be found by taking the derivative! The series we get for sine and cosine converge very quickly, and so they furnish excellent numerical approximations. This is true of the series for e^x as well.

From a more advanced perspective, the series for $\sin(x)$ and $\cos(x)$ serve as the best way to define them. It may come as a surprise that the key properties of these functions follow very directly from the series. We mentioned in the Introduction that the use of series furnishes a number of advantages.

Series for the logarithm can be found by integration. The geometric series

$$\frac{1}{1-t} = \sum_{n=0}^{\infty} t^n$$

has radius of convergence equal to 1, and so if $|x| < 1$, then Theorem 8.8 justifies

$$-\ln(1-x) = \int_0^x \frac{dt}{1-t} = \int_0^x \left(\sum_{n=0}^{\infty} t^n\right) dt = \sum_{n=0}^{\infty} \frac{x^{n+1}}{n+1}$$

When $x = -1$, the series is the convergent *alternating harmonic series*, and so, by Abel's Theorem (Theorem 8.10), we have

$$-\ln 2 = \lim_{x \to -1-} -\ln(1-x) = \lim_{x \to -1-} \sum_{n=0}^{\infty} \frac{x^{n+1}}{n+1} = \sum_{n=0}^{\infty} \frac{(-1)^{n+1}}{n+1}$$

which shows that $\ln(2)$ is the sum of the alternating harmonic series.

$$\ln(2) = 1 - \frac{1}{2} + \frac{1}{3} - \frac{1}{4} + \cdots$$

If $-x$ is substituted in for x in $\ln(1-x)$, we obtain a series formula for $\ln(1+x)$ with radius 1.

6. Problems

1. Show that a sequence that is eventually 0 has radius of convergence infinity. (Note: this applies, for example, for $a_k = \binom{n}{k}$ where n is a given non-negative integer.)

2. Let a_n, for $n \geq 0$, be a positive sequence, let $r > 0$, and suppose that $a_n/a_{n+1} \geq r$ for all n. Show that $a_n \cdot r^n$ is bounded. (Hint: show that $a_n \cdot r^n$ is decreasing.)

3. The *Fibonacci sequence* is defined by $a_0 = a_1 = 1$ and $a_{n+2} = a_{n+1} + a_n$ when $n \geq 0$. Use the previous problem to show that the radius of convergence of the Fibonacci sequence is at least $1/2$.

6. PROBLEMS

4. Find the radius of convergence of the following famous sequences.

a) $\dfrac{1}{n}$ b) 2^n c) $\dfrac{1}{n!}$ d) $\dfrac{1}{(2n)!}$

5. Let α be a real number that is not a non-negative integer. Show that the sequence $a_k = \binom{\alpha}{k}$ has radius of convergence 1. (To repeat the hypothesis: either α is not an integer at all, or α is a negative integer. Hint: use recursion.)

6. Apply the Corollary and Proposition 8.2c to show that if a_n is a polynomial in n, then the radius of convergence is 1.

7. This problem illustrates the use of *generating functions* in combinatorics and computer science. Recall the Fibonacci sequence a_n for $n \geq 0$; you have shown that its radius of convergence R is at least $1/2$. Thus,

$$f(x) = \sum_{n=0}^{\infty} a_n \cdot x^n$$

is defined for $|x| < 1/2$. Show that $f(x) = 1 + x \cdot f(x) + x^2 \cdot f(x)$. Use that equation to solve for $f(x)$ as a rational function! (Note: a Taylor series for the rational function can be found by other means, and an algebraic formula for the Fibonacci sequence results.)

8. If $y \in [0,1]$ and if n, m are non-negative integers with $n \leq m$, then

(8.6) $$\sum_{k=n+1}^{m} |y^k - y^{k-1}| \leq 1$$

(Hint: the sum collapses.)

9. Show that the function $(1+x)/(1-x)$ maps the open interval $(-1, 1)$ one to one, onto the positive real numbers. Use the series for $\ln(1+x)$ and $\ln(1-x)$ to get a series for $\ln((1+x)/(1-x))$. (Note: this latter series can be used to calculate the logarithm of an arbitrary positive number.)

10. Here are some formulas which show up in combinatorics. Let n, k be non-negative integers, and show that

$$\frac{1}{(1-x)^n} = \sum_{k=0}^{\infty} \binom{k+n-1}{k} \cdot x^k$$

Find the radius of convergence of these formulas, as well.

11. In the previous problem you obtained a likely series for $1/(1-x)^2$. Obtain the (famous!) numerical formula obtained from this series at $x = 1/2$.

12. From the Taylor series for $\cos(x)$, show that $\cos(2) < 0$. (Hint: the sum of the first three terms is negative; after that, each pair of terms adds to something negative.)

13. (Continuing the previous problem.) Since $\cos(0) > 0 > \cos(2)$, there is a minimum number α in $[0, 2]$ such that $\cos(\alpha) = 0$. (Why?) Define $\pi = 2 \cdot \alpha$. Show that $\sin(\pi/2) = 1$. (Hint: not hard to show that $\sin^2(\pi/2) = 1$. Is $\sin(x)$ increasing or decreasing on $[0, \pi/2]$?)

14. (Continuing the previous problem.) Prove that $\cos(\pi) = -1$ and $\sin(\pi) = 0$, and that $\cos(2 \cdot \pi) = 1$ and $\sin(2 \cdot \pi) = 0$. Show that $\cos(x)$ and $\sin(x)$ have period $2 \cdot \pi$. (Hint: A problem on p.79 gives you the angle addition equations; for instance, you have the double angle formulas.)

15. Find a series for $\arctan(x)$, using its integral formula from a problem on p.116. (Hint: the integrand is the sum of a geometric series.)

16. Find a Taylor series about 0 for a function $f(x)$ such that $f''(x) = f(x)$ and $f(0) = 0$ and $f'(0) = 1$. What is the radius of convergence? (Note: this function is called the *hyperbolic sine* and denoted $\sinh(x)$; its derivative is the *hyperbolic cosine* $\cosh(x)$.)

17. Use the differential equation in the previous problem to prove that $\cosh^2(x) - \sinh^2(x) = 1$. (Hint: take the derivative of the left side.)

18. Here is another use of series to solve a differential equation. Assume that y is a Taylor series about 0, where

$$y'' - x^2 \cdot y = x \quad \text{and} \quad y(0) = y'(0) = 0$$

Use the differential equation to solve for the coefficients. Show that the radius of convergence of those coefficients is infinity, and then show that the series does indeed satisfy the differential equation.

19. It is of great interest in statistics to calculate

$$\int_0^x \exp(-t^2) \cdot dt$$

Find a Taylor series in x for this integral.

20. Continuing the previous problem: for $|x| < 10$, use the infinite series for the exponential function and Taylor's remainder formula to find a finite sum approximation within 10^{-10} of the integral.

Note You may have seen the complex number formulas that relate e^x to $\cos(x)$ and $\sin(x)$. Although complex numbers do not play a formal role in our course, we will play with them in the following interesting and suggestive exercises.

21. Suppose that $i^2 = -1$. Use $i \cdot x$ in place of x in the Taylor series for e^x and obtain *Euler's Identity*:

$$\exp(i \cdot x) = \cos(x) + i \cdot \sin(x)$$

22. Prove the following. As in the previous problem: $i^2 = -1$.
a) $\exp(-i \cdot x) = \cos(x) - i \cdot \sin(x)$
b) $\cos(x) = (\exp(i \cdot x) + \exp(-i \cdot x))/2$.
c) $\cosh(x) = \cos(i \cdot x)$

APPENDIX A

The Real Numbers.

The following is meant to accompany our discussion in class about the nature of the real numbers. We begin with the arithmetic and order properties with which you are already familiar.

1. Properties Except for Completeness.

We categorize the properties as arithmetic properties (labeled **A** in the Axioms) and order properties (labeled **O**).

Real Number Axioms: There is a set $\mathbb{R}$ such that $\mathbb{Q} \subset \mathbb{R}$, and such that for all $a, b \in \mathbb{R}$, there are $a + b, a \cdot b \in \mathbb{R}$, and also the statement $a < b$ which is true or false. We also have

A1. If $x, y \in \mathbb{R}$, then $x + y = y + x$.
A2. If $x, y, z \in \mathbb{R}$, then $(x + y) + z = x + (y + z)$.
A3. If $x \in \mathbb{R}$, then $x + 0 = x$ and $x \cdot 0 = 0$. ($0 \in \mathbb{Q}$)
A4. If $x \in \mathbb{R}$, there is a unique $y \in \mathbb{R}$ such that $x + y = 0$.
A5. If $x, y \in \mathbb{R}$, then $x \cdot y = y \cdot x$.
A6. If $x, y, z \in \mathbb{R}$, then $(x \cdot y) \cdot z = x \cdot (y \cdot z)$.
A7. If $x \in \mathbb{R}$, then $x \cdot 1 = x$. ($1 \in \mathbb{Q}$)
A8. If $x \in \mathbb{R}$ and $x \neq 0$, then there is a unique $y \in \mathbb{R}$ such that $x \cdot y = 1$.
A9. If $x, y, z \in \mathbb{R}$, then $x \cdot (y + z) = (x \cdot y) + (x \cdot z)$.
O1. If $x, y \in \mathbb{Q}$, then $x < y$ in $\mathbb{R}$ if and only if $x < y$ in $\mathbb{Q}$.
O2. If $x, y \in \mathbb{R}$, then exactly one of the following holds: $x < y$, $x = y$, $y < x$.
O3. If $x, y, z \in \mathbb{R}$ and $x < y$ and $y < z$, then $x < z$.

O4. If $x, y, z \in \mathbb{R}$ and $x < y$, then $x + z < y + z$.

O5. If $x, y, z \in \mathbb{R}$ and $x < y$ and $0 < z$, then $(x \cdot z) < (y \cdot z)$. If $z < 0$, then $(y \cdot z) < (x \cdot z)$. ∎

The unique additive inverse for x identified in A4 as y is denoted $-x$, as you expect, and it follows from A3 that $-x = (-1) \cdot x$. The unique multiplicative inverse for x identified in A8 is denoted $1/x$ or x^{-1}, as usual.

Of course, the *positive* real numbers are those $x \in \mathbb{R}$ with $0 < x$, and the *negative* real numbers are those $x \in \mathbb{R}$ with $x < 0$. We will write $x > y$ for $y < x$, we will write $x \leq y$ for "$x < y$ or $x = y$," and we will observe all the other common inequality notations.

We will assume that you are familiar with the exponential notation: x^n where $x \in \mathbb{R}$ and n is a positive integer. And we will assume that you know the basics about polynomials, as well; for instance you should know that roots arise from linear factors.

The familiar open and closed intervals: for real numbers a, b with $a < b$, we define the *open interval* (a, b) to be the set of $x \in \mathbb{R}$ such that $a < x < b$. You know that the notation (a, b) sometimes denotes an *ordered pair*, for instance as a point in the plane. In this course, (a, b) will always be an open interval. We also allow the open interval (a, ∞) consisting of all real numbers $x > a$, the open interval $(-\infty, a)$ of all $x < a$, and the entire set of real numbers $\mathbb{R} = (-\infty, \infty)$.

If $a, b \in \mathbb{R}$ and $a \leq b$, then the *closed interval* $[a, b]$ is the set of $x \in \mathbb{R}$ with $a \leq x \leq b$. And the *half-open interval* $(a, b]$ consists of x with $a < x \leq b$. Similarly, $[a, b)$ has the obvious meaning.

We also assume you are familiar with the absolute value and its elementary properties. The Triangle Inequalities are given as exercises in Chapter 1.

2. Construction.

Now we show how to construct the real numbers from the rational numbers via *Dedekind cuts*. Our construction produces the properties we just listed, along with the Completeness Property. Dedekind's idea appears in one of his books, translated into English as *Essays on the Theory of Numbers,* published by Dover in 1963. This section is included to set the course material in a more complete framework; it will not be used in class.

There are, actually, several ways to *construct* the real numbers, so it might be good to say something briefly about this issue. The Axioms above postulate the *existence* of a set $\mathbb{R}$. Some people feel that the set of real numbers are *known* to exist without a formal construction – for instance, they might be understood as the set of signed lengths. There are, in fact, several possible opinions about this. The purpose of this course is more mathematical than philosophical, and so we will emphasize the mathematical details. Since we are doing mathematics within the set theory and its logic, as introduced in math 300, we will work within that framework – this leaves open the question of the relation between our construction and other ways to construct and/or understand the real numbers.

We will assume that the rational numbers have been constructed, and that they satisfy the Axioms given above: A1-A9, O1-O5. We will also use the *Archimedean Property* that the integers are not bounded above and not bounded below in the rationals. Dedekind's idea is to represent the real number c via the set $\mathbb{Q} \cap (-\infty, c)$. This set consists only of rational numbers – previously defined. The trick is to characterize this set without mentioning the number c directly, so that the set can be used to *define* c. Here goes. A *Dedekind cut* (*cut* for short) is a set C where

(a) C is a non-empty, proper subset of $\mathbb{Q}$
(b) if $x < y \in C$, then $x \in C$
(c) the set C does not have a maximum

Remember that a cut is a subset of the rational numbers! Here is a preliminary[1] definition of the real numbers: a real number is a cut, and the set of real numbers is the set of cuts. For the rest of this chapter we will gradually develop the necessary properties listed above. Throughout our arguments, lower case letters will stand for rational numbers.

We begin with the cut that will represent the rational number r.

1. Rational cuts. For each $r \in \mathbb{Q}$, define ∂r to be the set of $s \in \mathbb{Q}$ with $s < r$. Then ∂r is a cut.[2] (Prove this!) Also, if $\partial r = \partial s$, then $r = s$. We will see that the set of rational cuts has the same properties as the set of rational numbers.

2. Approximation. The following technical lemma will be very useful. As is common with technicalities, it is better appreciated *after* you have absorbed the rest of these notes.

APPROXIMATION LEMMA. *Let $e \in \mathbb{Q}$ with $e > 0$, and suppose that C is a cut. Then there is $c \in C$ with $c + e \notin C$.*

PROOF. By (a), there is $x \in C$, and $y \in \mathbb{Q} \setminus C$. By (b) it must be that $x < y$. There is a positive integer $k \geq (y-x)/e$, and so $x + ke \geq y$, and therefore $x + ke \notin C$. Thus, the set of positive integers j such that $x + je \notin C$ is not empty; by Well-ordering it has a minimal element j. Then $x + je \notin C$, but by the minimality of j, we must have $x + (j-1)e \in C$. Put $c = x + (j-1)e$ and we have what we wanted. □

3. Comparison. Let C, D be cuts and suppose that C is not a subset of D. Then there is $x \in C$ with $x \notin D$. If $d \in D$, then (b) applied to D shows

---

[1] The definition will need to be modified slightly to accommodate the rational numbers; we will discuss this later.

[2] We have worried over what symbol to use for a rational cut. The use of ∂r is not perfectly satisfactory, but at least it's easy to write.

that $d < x$, and (b) applied to C then shows that $d \in C$. Thus, $D \subseteq C$. This proves that if C, D are cuts, then exactly one of the following is true: $C \subset D$ or $C = D$ or $D \subset C$.

We define $C < D$ to mean $C \subset D$, and we obtain an *ordering* that satisfies Axioms O1,O2,O3. You should prove for all r, s, that $r < s$ if and only if $\partial r < \partial s$.

4. Addition. For cuts C, D, define $C + D$ to be the set of all $c + d$ with $c \in C$ and $d \in D$. It is easy to show that $C + D$ is a cut: it is clearly not empty. There is $x \in \mathbb{Q} \setminus C$, and (b) shows that $x > c$ for all $c \in C$. Similarly, there is $y \in \mathbb{Q} \setminus D$ and the $y > d$ for all $d \in D$. It follows that $x + y > c + d$ for all $c \in C$ and $d \in D$, and this shows that $x + y \notin C + D$. We have proved (a) for $C + D$.

If $x < c + d$ where $c \in C$ and $d \in D$, then $x - c < d$, so that $x - c \in D$. Now $x = c + (x - c)$ shows that $x \in C + D$. Thus (b) holds for $C + D$. Again if $c \in C$ and $d \in D$, then (c) finds $c' \in C$ with $c' > c$ and the same property finds $d' \in D$ with $d' > d$. Then $c' + d' \in C + D$ and $c' + d' > c + d$, so that (c) holds for $C + D$.

We leave it to the reader to show that addition of cuts obeys the commutative and associative laws – Axioms A1,A2. You should also show that $\partial r + \partial s = \partial(r + s)$ for all r, s. Thus, addition on the reals extends addition on the rational cuts.

We claim that $\partial 0$ is the additive identity, so that A3 holds. Indeed, if C is a cut and $c \in C$, then (c) finds $c' \in C$ with $c' > c$. Then $c - c' \in \partial 0$, and so $c = (c - c') + c'$ represents c as an element of $\partial 0 + C$. This proves that $C \leq \partial 0 + C$. For the converse, if $r \in \partial 0$ and $c \in C$, then $r < 0$, so that $r + c < c$ implies that $r + c \in C$, as needed.

Now we can prove the existence of additive inverses: Given the cut C, there is a cut D such that $C + D = \partial 0$. Intuitively, we would like to define D as the set of $-c$ for $c \in C$, but this does not define a cut (it does not have an upper bound!), so we need to be trickier. Let D be the set of x such that there is $y \notin C$ with $y < -x$. We will show that D is a cut; the proof is not really hard, we just have to make careful use of the definition of D. By (a) applied to C, there is $y \notin C$. Choose $x > y$, and then $-x \in D$. Thus D is non-empty. Let $r \in C$ and we claim that $x < -r$ for all $x \in D$. Indeed, if $-r \leq x$, then $-x \leq r$ implies that $-x \in C$. But there is $y \notin C$ with $y < -x$, and this is a contradiction. Thus, (a) holds for D. If $x \in D$ and $z < x$, then there is $y \notin C$ with $y < -x$, and we see that $y < -z$ as well, so that $z \in D$. So (b) holds for D. If $x \in D$, there is $y \notin C$ with $y < -x$. There is $y < z < -x$, and then $-z \in D$ and $x < -z$; so (c) holds. We have proved that D is a cut.

Now we prove that $C + D \leq \partial 0$. If $c \in C$ and $d \in D$, there is $y \notin C$ with $y < -d$. It follows that $c < y$, and so $c + d < 0$, whence $c + d \in \partial 0$ as claimed.

Next: $\partial 0 \leq C + D$. Let $r \in \partial 0$, so that $r < 0$. Noting that $-r/2 > 0$, the Approximation Lemma finds $c \in C$ such that $c - r/2 \notin C$. Also, we have $c - r/2 < c - r$. By definition of D, we have $r - c \in D$. Now $r = c + (r - c)$ shows that $r \in C + D$.

We have proved the existence statement in Axiom A4; the uniqueness follows easily using the associative law. Each cut C now has a unique negative $-C$.

We have Axioms A1,A2,A3,A4,O1,O2,O3. Next we prove O4. Let C, D, E be cuts with $C < D$, and we will prove that $C + E < D + E$. An element of $C + E$ can be written $c + e$ where $c \in C$ and $e \in E$. Since $C \subset D$, we have $c \in D$, and so $c + e \in D + E$. This proves that $C + E \subseteq D + E$. If $C + E = D + E$, then adding $-E$ shows that $C = D$, a contradiction. Thus, $C + E < D + E$.

5. Completeness. The most elegant thing about Dedekind's construction is how easily it establishes Completeness. For let S be a non-empty set of cuts that is bounded above. We claim that the union of the elements of S is a cut, and it is the least upper bound of S.

PROOF. Let T be the union of the elements of S. It is easy to see that T satisfies (b) and (c). It is also obvious that T is not empty. To complete the proof that T is a cut, we need to show that it is proper. There is a cut C with $C \geq D$ for all $D \in S$. Let $x \notin C$ and $t \in T$. Then $t \in D$ for some $D \in S$. If $x \leq t$, then $x \in D$, so that $x \in C$, a contradiction. Thus, $t < x$ for all $t \in T$, and so $x \notin T$, as needed.

The definition of T shows that $D \leq T$ for all $D \in S$, and so T is an upper bound. If U is an upper bound for S, then $D \subseteq U$ for all $D \in S$, so that $T \subseteq U$. In other words, $T \leq U$, and we see that T is the least upper bound. $\square$

Since we have addition, subtraction, and comparison, we can define the limit of a sequence of cuts as done in this text in the chapter on sequences. Completeness then gives us the Monotone Convergence Theorem. We will show how to use this theorem to extend multiplication from the rationals to the reals. We need the following.

PROPOSITION A.1. *Let C be a cut. Then C is the limit of an increasing, constant-sign sequence ∂c_n.*

PROOF. Use the Approximation Lemma to find $r_1 \in C$ with $r_1 + 1 \notin C$. If $C > \partial 0$, we can choose $r_1 > 0$. If $C \leq \partial 0$, then $r_1 < 0$. Given $r_1 < r_2 < \cdots < r_n \in C$ with $r_n + 1/n \notin C$, the Approximation Lemma finds $r_{n+1} \in C$ with $r_{n+1} + 1/(n+1) \notin C$. By cut property (c) we can replace r_{n+1}, if necessary, by a rational greater than r_n. It is easy to see that $\partial r_n \to C$ as $n \to \infty$ and that the r_n have the same sign. $\square$

6. Multiplication. Let C, D be cuts. Proposition A.1 finds constant-sign increasing sequences ∂c_n and ∂d_n that converge to C, D, respectively. It follows that the sequence $c_n \cdot d_n$ is monotone as well. We intend to define $C * D$ to be the limit of $\partial(c_n \cdot d_n)$. First we need to notice that $c_n \cdot d_n$ is bounded, since the c_n and d_n are bounded; thus, the Monotone Convergence Theorem gives the product sequence a limit. Second, we need to show that the product sequence limit does not depend on the particular sequences c_n, d_n. Indeed, suppose that e_n is a constant-sign increasing sequence such that $\partial e_n \to C$. Then

$$\partial(e_n \cdot d_n) = \partial(c_n \cdot d_n + (e_n - c_n) \cdot d_n) = \partial(c_n \cdot d_n) + \partial((e_n - c_n) \cdot d_n)$$

Since $\partial(e_n - c_n) \to C - C = \partial 0$, it is easy to prove that $\partial((e_n - c_n) \cdot d_n) \to \partial 0$, and then

$$\lim_{n \to \infty} \partial(e_n \cdot d_n) = \lim_{n \to \infty} \partial(c_n \cdot d_n)$$

These considerations show that we can define $C * D$ to be the limit of $\partial(c_n \cdot d_n)$ where c_n and d_n are sign-constant monotone sequences such that $\partial c_n \to C$ and $\partial d_n \to D$.

The properties of multiplication follow easily: Axioms A5,A6,A7,A8,A9. For instance, we can show that $C * (D + E) = (C * D) + (C * E)$ in the case that D and E have the same sign. Indeed, let $\partial c_n \to C$ and $\partial d_n \to D$ and $\partial e_n \to E$, as above. Then d_n and e_n have the same sign, and so $d_n + e_n$ is a constant-sign increasing sequence that converges to $D + E$. Thus,

$$C * (D + E) = \lim_{n \to \infty} \partial(c_n \cdot (d_n + e_n))$$
$$= \lim_{n \to \infty} \left(\partial(c_n \cdot d_n) + \partial(c_n \cdot e_n) \right) = C * D + C * E$$

as needed.

This completes our construction except for one detail: we need the rational numbers to be a subset of the real numbers. We have seen that the set ∂r behaves exactly as do the rational numbers r. We can *identify* r and ∂r – that means we agree never more to distinguish them.

APPENDIX B

Equivalent to Completeness.

Once we had the arithmetic and order axioms for the reals, we stated the Completeness Property as an Axiom and we derived Monotone Convergence from that. We then derived BW from Monotone Convergence. You should check carefully that the latter proof did not use Completeness directly – only the arithmetic and order axioms for the reals along with Monotone Convergence. In other words, if Monotone Convergence had been substituted for Completeness as an axiom, we would still have been able to derive BW. It turns out that we could have started with BW and derived Monotone Convergence and Completeness from it. In fact, there are many equivalent formulations of the completeness axiom; it is the purpose of this section to exhibit some of them.

We derived the Archimedean Property of the reals from completeness. It turns out that it can be derived from many of the equivalents to completeness as well. Adding the proof of this makes many of our arguments unwieldy, and so we choose to assume the Archimedean Property from the get-go. This is not too much of a compromise: the Archimedean Property was identified hundreds of years before the real numbers were formalized.

We defined what it means for a number to be a limit point of a set of real numbers. We will need the related but distinct concept of the limit point of a *sequence*. For a sequence a_n, the number p is a *limit point* of the sequence if for all $\epsilon > 0$ there are infinitely many values of n such that $|a_n - p| < \epsilon$.

We also need to define the *open* subsets of the real numbers. The set $V \subseteq \mathbb{R}$ is open if for every $x \in V$, there is $r > 0$ such that $(x-r, x+r) \subseteq V$. Another way to say this: an open set is a union of open intervals. A set $C \subseteq \mathbb{R}$ is *closed* if $\mathbb{R} - C$ is open. These definitions are the starting point of the subject of *topology*. In this book, we merely introduce them to make sense of the following.

THEOREM B.1. *Assume the arithmetic and order axioms for the real numbers listed in Appendix A, and assume the Archimedean Property. Then the following are equivalent.*

1. *Every non-empty subset of the reals that is bounded above has a least upper bound.*

2. *If I_j is a sequence[1] of closed intervals such that $I_{n+1} \subseteq I_n$ for all $n \geq 1$, then there is a real number in all the I_j.*

3. *(BW for sets) If S is a non-empty, bounded, infinite subset of the reals, then S has a limit point.*

4. *(BW for sequences) Every bounded sequence of real numbers has a sequence limit point.*

5. *Every bounded sequence of real numbers has a converging subsequence.*

6. *If I_j is a sequence of closed, bounded subsets of the reals and if the intersection of all the I_j is empty, then the intersection of finitely many of them is empty.*

7. *If C is a closed, bounded subset of the reals, and if V_j is a sequence of open subsets of the reals such that C is contained in the union of all the V_j, then C is contained in the union of finitely many of the V_j.*

8. *(Monotone Convergence) Every bounded monotone sequence has a limit.*

---

[1] For this proof, a sequence has the positive integers as domain.

9.(*Intermediate Value Theorem*) *If $f(x)$ is continuous on the closed interval $[a,b]$, and if N is a number between $f(a)$ and $f(b)$, then there is $c \in [a,b]$ with $f(c) = N$.*

PROOF. (1)$\Rightarrow$(2). Let a_j be the left endpoint of I_j; they are bounded above by the right endpoint of I_1. Let s be the sup of the a_j, and then $s \geq a_j$ for all j. Suppose that $s \notin I_n$, and let b be the right endpoint of I_n. Then $b < s$, and the interval $(b, s]$ contains no a_j, a contradiction.

(2)$\Rightarrow$(3). Let $S \subseteq [a_0, b_0]$. Given the closed interval I_j such that $S \cap I_j$ is infinite, choose the left or right half of I_j to be I_{j+1} based on the same property. There is a point s in the intersection of all the I_j. Let $\epsilon > 0$. The Archimedean Property can be used to show that there is j such that $I_j \subseteq (s - \epsilon, s + \epsilon)$, and so the open interval contains infinitely many elements of S. Thus, s is a limit point of S.

(3)$\Rightarrow$(4). If the sequence has infinitely many values, then the set of values has a limit point, which is a sequence limit point. If the sequence has finitely many values, then one of its values repeats infinitely often and is a sequence limit point.

(4)$\Rightarrow$(5). Let p be a sequence limit point for the bounded sequence a_j. Choose $a_{f(1)} \in (p-1, p+1)$. For each $n \geq 2$, choose $a_{f(n)}$ such that $f(n)$ is greater than $f(m)$ for all $m < n$ and such that $a_{f(n)}$ is within $1/n$ of p. Then $a_{f(n)} \to p$ as $n \to \infty$.

(5)$\Rightarrow$(6). The contrapositive: suppose every finite intersection is non-empty. Let $a_j \in \bigcap_{i=1}^{j} I_i$, for each j. Each I_1 is bounded, and so the a_j are bounded. By (5) there is a converging subsequence, let its limit be p. If $p \notin I_j$, then there is $\epsilon > 0$ such that $(p - \epsilon, p + \epsilon) \cap I_j = \phi$. No a_k with $k \geq j$ can be in this interval, and this is a contradiction. Thus, p is in all the I_j.

(6)$\Rightarrow$(7). Define $I_j = C \setminus V_j$, so that I_j is closed and bounded. The intersection of all the I_j is empty, since the union of all the V_j contains C.

By (6) there is a finite subset F of the positive integers such that the intersection of I_j with $j \in F$ is empty. It follows that the union of V_j for $j \in F$ contains C.

(7)$\Rightarrow$(8). Let a_n be an increasing sequence, and let b be an upper bound. (We will leave the case of a decreasing sequence to the reader.) Suppose that for each $x \in [a_1, b]$, there is $\epsilon > 0$ such that $(x - \epsilon, x + \epsilon)$ contains a_n for only finitely many values of n. The statement (7) implies that $[a_1, b]$ is the union of finitely many such open intervals, and it follows that a_n exists for only finitely many values of n. This is a contradiction, and so there is $x \in [a_1, b]$ such that $(x - \epsilon, x + \epsilon)$ contains a_n for infinitely many values of n, for every $\epsilon > 0$. Because a_n is increasing, it follows that $a_n \to x$ as $n \to \infty$.

(8)$\Rightarrow$(9). The bisection algorithm produces two sequences:
$$a = a_0 \leq a_1 \leq a_2 \leq \cdots \leq b_2 \leq b_1 \leq b_0 = b$$
such that, for each j, the number N is between $f(a_j)$ and $f(b_j)$. Also, each $b_{j+1} - a_{j+1}$ is half as big as $b_j - a_j$, and it follows from the Archimedean Property that $b_j - a_j \to 0$ as $j \to \infty$. By (8), the sequences a_j and b_j converge, and since $a_j - b_j \to 0$, the two sequences converge to the same number p. Because f is continuous, $f(a_j) \to f(p)$ and $f(b_j) \to f(p)$, and this shows that $f(p) = N$.

(9)$\Rightarrow$(1). Let S be a non-empty subset of the reals that is bounded above but has no least upper bound. Let B be the set of upper bounds for S, and define $f : \mathbb{R} \to \mathbb{R}$ by setting $f(x) = 1$ when $x \in B$, and $f(x) = 0$ when $x \notin B$. We claim that f is continuous on the reals. If $x \in B$, then since x is not least, there is $y \in B$ with $y < x$. Then f is constant 1 on (y, ∞); this proves that f is continuous at x. If $x \notin B$, then there is $y \in S$ such that $x < y$. Then f is constant 0 on $(-\infty, y)$, and f is continuous at x. Property (9) shows that $1/2$ is in the image of f, and this is a contradiction. Thus, S has a least upper bound. □

APPENDIX C

Counting.

1. Finite Sets.

We begin this section with a review of elementary facts. We want to make sure we are all on the same page. We will move toward a more advanced fact: that we can enumerate the rational numbers but not the real numbers.

A set S is *finite* if it is empty or if there is a positive integer n and a one to one, onto function $f : \{1, 2, \ldots, n\} \to S$. The number n is sometimes called the *order* of S. We say that the empty set has *order 0* (it has 0 elements).

PROPOSITION C.1. *A finite set S has a unique order. If S and T are finite sets, then $S \cup T$ is finite. If they are disjoint, then the order of $S \cup T$ is the sum of the order of S and the order of T.*

We also have

THE PIGEON-HOLE PRINCIPLE. *If T is a subset of the finite set S, then T is finite and its order is less than or equal to the order of S, and if the order of T is equal to the order of S, then $T = S$.*

You have proved that every non-empty, finite subset of the reals has a maximum and minimum, and that such a set can be put into numerical order.

2. Countable Sets.

This topic is a little off the path, but it is commonly included in beginning analysis courses. Let N denote the set of positive integers. A set C is *countable*[1] if there is a one to one function $f : C \to N$. We also allow the empty set to be countable. We will mention in class how the notion of countability arose out of the desire to compare the sizes of infinite sets.

Some infinite sets are countable (an obvious example: the positive integers themselves), and some aren't. An interesting difference between the rational numbers and the real numbers is disclosed in the fact that the rational numbers *are* countable, whereas the real numbers *are not*. Both of these facts are surprising: for one thing, it would seem that there are *more* rational numbers than positive integers; for another, given that that rationals and integers are both countable, it is surprising that the real numbers are not countable since the rational numbers are dense.

Here is a proof that the rationals are countable.

*PROPOSITION C.2. *The set $\mathbb{Q}$ is countable.*

PROOF. Each rational number can be written in the form $s \cdot a/b$ where $s = \pm 1$, and a is a non-negative integer, and b is a positive integer. Define

$$f(s \cdot a/b) = 2^{s+1} \cdot 3^a \cdot 5^b$$

Notice that $f(s \cdot a/b)$ is a positive integer. By uniqueness of prime factorization in the integers, the numbers a, b, c can be recovered from the positive integer $2^{s+1} \cdot 3^a \cdot 5^b$, and this shows that $f : \mathbb{Q} \to N$ is one to one. □

Here is a lemma that will move us further along.

LEMMA C.3. *Let C be an infinite subset of N. Then there is a one to one, onto function $h : C \to N$.*

---

[1] There are many equivalent definitions of *countable*.

PROOF. For $c \in C$, note that $C \cap [1, c]$ is a finite set. Define
$$h(c) = |C \cap [1, c]|$$

To show that h is one to one, let $a \leq b$ be elements of C and suppose that $h(a) = h(b)$. Notice that
$$C \cap [1, a] \subseteq C \cap [1, b]$$
and these two finite sets have the same number of elements, and so they are the same set. In particular, $b \in C \cap [1, a]$ shows that $b \leq a$, so that $a = b$.

To show that h is onto, we will prove $h(C) = N$ using induction. Let c be the minimal elements of C. Then $C \cap [1, c] = \{c\}$, and so $h(c) = 1$. Thus, $1 \in h(C)$.

Assume that $k \in h(C)$, and let $h(a) = k$. The set $C \cap [1, a]$ is finite, and since C is infinite, the set $C \cap [a+1, \infty)$ is not empty. This set has a minimal element b. We see that
$$C \cap [1, b] = \Big(C \cap [1, a]\Big) \cup \{b\}$$
and so $h(b) = h(a) + 1 = k + 1$. Thus, $k + 1 \in h(C)$. This proves that $h(C) = N$. □

In our definition of countable, we made the function *one to one*; if the set is infinite, the function can be made onto as well. This fact is sometimes taken as the definition of countable for infinite sets.

PROPOSITION C.4. *Let C be an infinite set. Then C is countable if and only if there is $g : N \to C$ that is one to one and onto.*

PROOF. Suppose that $g : N \to C$ is one to one and onto. Then $g^{-1} : C \to N$ is one to one, and so C is countable.

Conversely, suppose that C is countable, and let $f : C \to N$ be one to one. Then $f(C)$ is an infinite subset of N, and Lemma C.3 finds $h : f(C) \to N$

that is one to one and onto. The function $h \cdot f : C \to N$ is one to one and onto and its inverse is our desired g. □

Taking Proposition C.2 and Proposition C.4 together, we see that there is a one to one, onto function $g : N \to \mathbb{Q}$. In other words,

$$g_1, \ g_2, \ g_3, \ \cdots$$

is a listing of each rational number exactly once. Such an *enumeration* of the rationals will be useful in constructing various examples in our course.

The definition of countable makes it obvious that every subset of a countable set is countable. Thus, if a, b are real numbers with $a < b$, then the set of rational numbers in the interval $[a, b]$ is a countable set. By Rational Density, this set is also infinite, and so, as above, Proposition C.4 produces a one to one, onto function $g : N \to [a, b] \cap \mathbb{Q}$.

The real numbers in a non-trivial closed interval are too numerous to be countable.

*PROPOSITION C.5. *Let $a < b$ be numbers. Then the set $[a, b]$ is not countable.*

PROOF. Let $g : N \to [a, b]$ and we will show that g cannot be onto. Proposition C.4 will show that $[a, b]$ is not countable. We will use a technique we have often used: a sequence of closed intervals will squeeze down on a point.

There is a closed interval I_1 contained in $[a, b]$ such that $g(1) \notin I_1$. We can also make sure that the width $|I_1|$ satisfies $0 < |I_1| \leq 1/2$. Next, there is a closed interval I_2 contained in I_1 such that $g(2) \notin I_2$ and $0 < |I_2| \leq 1/4$. We can keep going: assume we have closed intervals

$$I_1 \supseteq I_2 \supseteq \cdots \supseteq I_n$$

and $0 < |I_n| < 1/2^n$. And we assume that $g(j) \notin I_j$ for each j with $1 \leq j \leq n$.

Now find a closed interval $I_{n+1} \subseteq I_n$ with $g(n+1) \notin I_{n+1}$ and $0 < |I_{n+1}| \leq 1/2^{n+1}$.

Proposition 2.7 in Chapter 2, applied to the endpoints of the I_n, finds a number c in all the I_n. We have that $c \in [a, b]$ and $c \neq g(n)$ for all n, since $c \in I_n$ and $g(n) \notin I_n$. This completes the proof. $\square$

3. Problems

1. Show that finite sets are countable.

2. Let C, D be countable sets, and let $f : C \to N$ and $g : D \to N$ be one to one. Define $h : C \cup D \to N$ as follows: for $x \in C \cup D$, if $x \in C$, define $h(x) = 2 \cdot f(x)$. Otherwise, $x \in D$, and define $h(x) = 2 \cdot g(x) + 1$. Show that h is one to one, so that $C \cup D$ is countable.

3. Let D be countable and let $g : C \to D$ be one to one. Show that C is countable.

Index

(a, b), open interval, 150
C^1 function, having a continuous derivative, 81
$[a, b]$, closed interval, 150
Δ-bound, 63
$\Sigma^u(f, P)$, upper Riemann sum, 96
$\Sigma_l(f, P)$, lower Riemann sum, 96
$\Sigma_l^u(f, P)$, the variation sum for f on P, 86
$\arcsin(x)$, inverse-sine, 76
$\arctan(x)$, inverse-tangent, 76
$\binom{\alpha}{n}$, the binomial sequence for α, 14
$\cos(x)$, cosine, 143
$\exp(x)$, construction, 113
$\inf(f, I)$, the inf of f on I, 96
$\int_a^b f$, definite integral of f over $[a, b]$, 99
$\lim_{k \to \infty} a_k$, the limit of the sequence a_k, 17
$\lim_{x \to \alpha} f(x)$, function limit, 40
$\ln(x)$, the natural logarithm, 112
$\mathbb{Q}$, the rationals, 2
$\mathbb{R}$, the reals, 2
$\mathbb{Z}$, the integers, 2
π, construction, 146

$\exp(x)$ or e^x, the exponential function, 10
$\sin(x)$, sine, 143
$\sum_{n=0}^{\infty} a_n$, the series for a_n, 127
$\sup(f, I)$, the sup of f on I, 95
$\mathrm{var}(f, I)$, the variation of f on I, 85
$|I|$, the length of the interval I, 86
$a_k \to A$, a_k goes to A, 16
e, Euler's number, 113
f', the derivative of f, 65
$f(x) \to A$, function limit, 37
$x \to a+$, limit from the right, 41
$x \to a-$, limit from the left, 41

absolute partial sums, 127
accumulation point, 31
alternating harmonic series, 144
arc length, 110
Archimedean Property, 5, 151

Babylonian sequence for the square root, 15
binomial sequence, 14
Binomial Theorem (arbitrary exponent), 142
Bolzano-Weierstrass Theorem, 32

bounded (sequence), 19
bounded above (sequence), 19
bounded above (set), 2
bounded below (sequence), 19
bounded below (set), 4
bounded set, 5
BW (Bolzano-Weierstrass Theorem), 32

Cauchy sequence, 34
Cauchy's MVT, 76
center (for Taylor series), 128
chain rule, 70
closed (subset of the reals), 158
closed interval, 1, 150
compact (closed intervals in the reals), 35
Completeness Property, 3
continuous (at a point), 49
continuous (on a domain), 51
Continuous IFT, 56
converges (sequence), 17
converges uniformly (on a domain), 119
cosine: construction, 143
countable, 162

decreasing (function), 74
decreasing (sequence), 23
Dedekind cut, 151
definite integral, 99
Delta-bound, 63
dense, 6
density of the rationals, 6
derivative, 65
differentiable, 65
differentiable (at a point), 65

Differentiable IFT, 74
Differentiable IVT, 81
Dirichlet function (for a sequence), 52
diverge (sequence), 17
Divergence Test, 127

Euler's identity for $\exp(i \cdot x)$, 147
Euler's number, 113
eventually (sequence property), 15
exponential function, 112
exponential function: construction, 113, 141
exponentials: arbitrary, 116
extreme point, 55
Extreme Value Theorem, 55

factorial sequence, 14
Fibonacci sequence, 144
finite (set), 161
Fundamental Theorem of Calculus, 105

generating functions, 145
greatest lower bound, 4

half-open interval, 1, 150
harmonic sequence, 14
Heine-Borel Theorem, on the reals, 35
hyperbolic cosine, 146
hyperbolic sine, 146

increasing (function), 73
increasing (sequence), 24
induction, 1
induction, review, 8
inf, 4

infimum, 4
infinite series, 127
infinity in limits, 44
infinity, as limit point, 45
infinity, what a concept, 44
injective (one to one), vi
integers, 1
integrable (on an interval), 86
integral function, 103
Integral MVT, 104
Intermediate Value Theorem, 52
irrational, 7
IVT (Intermediate Value Theorem), 52

L'Hôpital's Rule, 76
least upper bound, 3
limit (of a function), 37
limit point, 31
limit point (sequence), 157
lower bound, 4
lower bound (sequence), 19
lower Riemann sum, 96

M Test (for uniform convergence), 120
MacLauren series, 129
maximum (set), 3
maximum of a function, 55
Mean Value Theorem, 72
mesh (of partition), 113
minimum, 5
minimum of a function, 55
Monotone Convergence Theorem, 24

natural logarithm, 112

negative, 150

one to one, vi
onto, vi
open (subset of the reals), 158
open interval, 1, 150
order (of a set), 161

partition, 86
Pascal's Triangle, 15
period (of a function), 64
pigeon-hole principle, 161
positive, 150
power rule (for derivatives), 70
power series, 125
product rule (for derivatives), 69

quotient rule (for derivatives), 69

radius of convergence, 129
Rational Density, 6
rational function, 51
rational numbers, 1
rationals are countable, 162
Real Number Axioms, 149
real numbers, 1
reals are not countable, 162
recursive definition, 14
refines (partition), 91
remainder (of Taylor series), 138
Riemann integral, 84
Riemann sum (arbitrary), 99

secant function, 66
second derivative, 71
sequence, 13
series (infinite), 127
sine: construction, 143

square root: construction, 26
strictly decreasing (function), 74
strictly decreasing (sequence), 23
strictly increasing (function), 74
strictly increasing (sequence), 24
subdivision, 86
sup, 3
supremum, 3
surjective (onto), vi

Taylor series, 129
Taylor's formula, 136
Taylor's Theorem with Remainder, 137
Theorem on Interior Extremes, 72
Triangle Inequality, 9
triangle inequality (for integrals), 102

uniform convergence, 119
uniformly continuous, 59
upper bound, 2
upper bound (sequence), 19
upper Riemann sum, 96

vacuous implication, 2
variation on a set, 85
variation sum, 86

Weierstrass' Comparison Test, 120
Well-Ordering (of the integers), 1

Made in the USA
Charleston, SC
04 March 2014